Solution Architecture with .NET

Learn solution architecture principles and design techniques to build modern .NET solutions

Jamil Hallal

BIRMINGHAM—MUMBAI

Solution Architecture with .NET

Group Product Manager: Aaron Lazar
Publishing Product Manager: Ashish Tiwari
Senior Editor: Nitee Shetty
Content Development Editor: Rosal Colaco
Technical Editor: Karan Solanki
Copy Editor: Safis Editing
Project Coordinator: Deeksha Thakkar
Proofreader: Safis Editing
Indexer: Tejal Daruwale Soni
Production Designer: Shankar Kalbhor

First published: July 2021

Production reference: 1230721

Published by Packt Publishing Ltd.
Livery Place
35 Livery Street
Birmingham
B3 2PB, UK.

ISBN 978-1-80107-562-6

www.packt.com

Contributors

About the author

Jamil Hallal is a results-driven .NET solutions architect with a strong track record of designing and developing enterprise software solutions that dramatically increase organizational effectiveness. He is certified as a Microsoft Certified Professional, designing and developing enterprise applications. He has extensive experience in building large-scale .NET web solutions, process automation, SharePoint portals, business intelligence and data analytics solutions, document management and archiving, microservices and service-based applications, and AI solutions, with a career spanning over 15 years in various industries.

About the reviewer

Gosia Borzecka is a Microsoft AI MVP and a modern workplace consultant at Avanade.

She is a full-stack .NET and React developer with Office 365 experience. For the last 2 years, she has been interested in AI and machine learning (and Python!), and in her day job, she helps customers bring AI and ML models into the modern workplace.

Gosia is also an international speaker and co-leader of NottsDevWorkshop, where she has organized a few AI/Office 365 Bootcamps. She also helps children learn about programming and new technology as a volunteer in local schools as a STEM Ambassador.

Table of Contents

3
What Makes an Effective Solution Architect?

Section 2: Designing a Solution Architecture

4
Designing a Solution Architecture

5

Exploring Architecture Design Patterns

6

Architecture Considerations

7

Securing ASP.NET Web Applications

8

Testing in Solution Architecture

Section 3: Architecting Modern Web Solutions with DevOps Solutions

9

Architecting Modern Web Solutions with ASP.NET Core and Azure

10
Designing and Implementing Microsoft DevOps Solutions

Other Books You May Enjoy

Index

Preface

In our rapidly evolving world, driven by digital transformation, solution architects are the most significant experts with a particular skill set and a wide range of technical expertise for balancing business needs with technology solutions. The purpose of this book is to give you a broad understanding of .NET solution architecture with a hands-on approach to help you become an effective solution architect.

The book covers the principles of the **software development life cycle** (**SDLC**), the roles and responsibilities of a .NET solution architect, and what makes a great .NET solution architect. As you progress through the chapters, you'll gain an understanding of the key principles of solution architecture and how to design a solution and explore designing layers and microservices.

You'll complete your learning journey by uncovering modern architecture patterns and techniques for designing and building digital solutions.

By the end of this book, you'll have learned how to architect your modern web solutions with ASP.NET Core and Microsoft Azure, and be ready to automate your development life cycle with Azure DevOps.

Who this book is for

This book is for intermediate and advanced .NET developers and software engineers who want to advance their careers and expand their knowledge of solution architecture and design principles. Beginner or intermediate-level solution architects looking for tips and tricks to build large-scale .NET solutions will also find this book useful.

What this book covers

Chapter 1, Principles of the Software Development Life Cycle, helps you understand that the concept and principles of the software development life cycle are a great kick-off point towards planning a software product. This chapter aims to explain the notion of SDLC, its phases, and methodologies.

Chapter 2, Team Roles and Responsibilities, focuses on the main roles in a typical software development team and their corresponding responsibilities. One of the key factors for a successful software project is to ensure that the key stakeholders of the development team are all in place. The success of the project also depends on how well the team works together.

Chapter 3, What Makes an Effective Solution Architect?, elaborates more on the personal qualifications needed to become a good solution architect.

Chapter 4, Designing a Solution Architecture, focuses on solution architecture practices by exploring the key principles of solution architecture and the most popular **Unified Modeling Language** (**UML**) diagrams that are recommended to design medium to large-scale solutions.

Chapter 5, Exploring Architecture Design Patterns, talks about modern architecture patterns with sample use cases. Additionally, we will explain the criteria that should be adopted to choose the right architecture pattern for our software solution.

Chapter 6, Architecture Considerations, explores design quality attributes and how to properly plan caching, exception handling, and deployment.

Chapter 7, Securing ASP.NET Web Applications, explores security considerations to be taken into account when designing a solution and looks at best practices in this context.

Chapter 8, Testing in Solution Architecture, explores different types of testing, including unit testing, stress testing, and automated testing.

Chapter 9, Architecting Modern Web Solutions with ASP.NET Core and Azure, helps you learn how to architect cross-platform modern web solutions with ASP.NET Core to best take advantage of its capabilities. Building web applications with ASP.NET Core, hosted in Azure, offers many competitive advantages over traditional alternatives. ASP.NET Core is optimized for modern web application development practices and cloud hosting scenarios.

Chapter 10, Designing and Implementing Microsoft DevOps Solution, helps you learn how to make use of Azure DevOps to build, test, and deploy applications by using modern software development practices. Moreover, we will get to know how to manage source control and we will also explore the management of packages using Azure Artifacts, as well as understanding Continuous Integration/Continuous Deployment practices.

To get the most out of this book

Here are a few requirements you should ensure are met before you start reading this book:

- You should be an intermediate or advanced .NET developer.

- You should have some basic knowledge of Microsoft Azure.

Software/hardware covered in the book	Operating system requirements
Visual Studio (latest version)	Windows, macOS, or Linux
Visio	Windows, macOS, or Linux

Download the color images

We also provide a PDF file that has color images of the screenshots and diagrams used in this book. You can download it here: `https://static.packt-cdn.com/downloads/9781801075626_ColorImages.pdf`.

Conventions used

There are a number of text conventions used throughout this book.

`Code in text`: Indicates code words in text, database table names, folder names, filenames, file extensions, pathnames, dummy URLs, user input, and Twitter handles. Here is an example: "It's quite easy to apply authorization in MVC by adding the `[Authorize]` attribute to the controller class or to the actions that are not anonymous."

A block of code is set as follows:

```
[Authorize(Users = "john,tim")]
public IActionResult EditContent()
{
  return View();
}
```

Any command-line input or output is written as follows:

```
Request URL:http://TheWebsiteUrl/register
Request Method:POST
Status Code:200 OK
firstname:John
```

Bold: Indicates a new term, an important word, or words that you see onscreen. For instance, words in menus or dialog boxes appear in **bold**. Here is an example: "As shown in the preceding screenshot, first we need to set the **Domain** name that we are using in the application."

> **Tips or important notes**
> Appear like this.

Get in touch

Feedback from our readers is always welcome.

General feedback: If you have questions about any aspect of this book, email us at customercare@packtpub.com and mention the book title in the subject of your message.

Errata: Although we have taken every care to ensure the accuracy of our content, mistakes do happen. If you have found a mistake in this book, we would be grateful if you would report this to us. Please visit www.packtpub.com/support/errata and fill in the form.

Piracy: If you come across any illegal copies of our works in any form on the internet, we would be grateful if you would provide us with the location address or website name. Please contact us at copyright@packt.com with a link to the material.

If you are interested in becoming an author: If there is a topic that you have expertise in and you are interested in either writing or contributing to a book, please visit authors.packtpub.com.

Share Your Thoughts

Once you've read *Solution Architecture with .NET*, we'd love to hear your thoughts! Scan the QR code below to go straight to the Amazon review page for this book and share your feedback.

https://packt.link/r/1-801-07562-X

Your review is important to us and the tech community and will help us make sure we're delivering excellent quality content.

Section 1: Understanding the Responsibilities of a Solution Architect

In this section, we will go through the different phases of the **Software Development Life Cycle (SDLC)** and we will learn about the differences between the popular SDLC models such as Scrum, Spiral, and DevOps. Then, we will learn about the hierarchy in a typical software development team and what to expect in terms of responsibilities from each member, including a solution architect.

Later in this section, we will explore some fundamental soft skills that every solution architect should have and we will get to know some common pitfalls that should be avoided.

This section comprises the following chapters:

- *Chapter 1, Principles of the Software Development Life Cycle*
- *Chapter 2, Team Roles and Responsibilities*
- *Chapter 3, What Makes an Effective Solution Architect?*

1
Principles of the Software Development Life Cycle

In the modern digital workplace, the role of the .NET solution architect is becoming crucial in the software development life cycle. Having a technology leader and a solution creator who can design and build robust and efficient solutions is a key factor to delivering successful products.

This book will highlight the fundamentals that you need to know, as a .NET professional developer, to become an effective solution architect in this growing and rapidly changing field.

Understanding the concept and principles of the **Software Development Life Cycle** (**SDLC**) is a great starting point toward planning a software product. This chapter aims to explain the notion of SDLC, its phases, and modern methodologies.

In this chapter, we will cover the following topics:

- Understanding what the SDLC is
- Exploring the different SDLC stages
- Getting familiar with the popular SDLC models

By the end of this chapter, you will be able to describe the SDLC stages and explain the difference between the popular SDLC models, such as **Scrum**, **Spiral**, and **DevOps**.

Understanding the software development life cycle

In today's digital world, every company is looking to deliver a good quality software product in a short period, which means the efficiency and the speed of the development team are game changers. To achieve this goal, companies must apply a set of well-defined activities and structured stages that define the **software development life cycle**, also known as the **SDLC**.

The SDLC is a methodology of work and best practices that aim to ease the process of software development and make it more efficient, ensuring the final product is delivered on time within the project budget and is totally in line with the expectations of the client.

There are different variations and models of the SDLC, such as the **Waterfall model**, the **Spiral model**, and the **Agile model**. They are popular and widely used by most software development organizations. Selecting the right model depends mainly on the size of the project and other factors. In the following sections, we are going to explore these models in detail to help you decide which model is right for your team and the project.

Here are the six stages that are defined in the SDLC process:

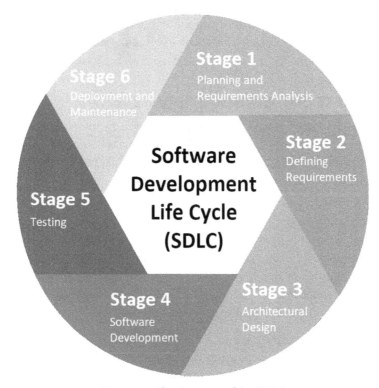

Figure 1.1: The six stages of the SDLC

We have just looked at an overview of the software development life cycle and its importance. In the next section, we will explore the different stages of the SDLC process.

Exploring the different SDLC stages

What are the main activities in the SDLC? No matter which model you choose to follow to implement your product, there are six different stages that are considered as common stages in most of the existing models. However, depending on the model, those stages can be executed sequentially or in parallel. By executing this series of stages, it is expected that you will be able to avoid typical and costly pitfalls and achieve the following goals:

- Lower costs
- Improved overall quality
- Shortened production time
- Excellent customer satisfaction

Let's explore these stages since understanding them is very important to the solution architect, who will be involved with all of them. On the other hand, knowing those stages is necessary to organize and facilitate the development of the product, as well as to make the entire development process more transparent. We'll understand each of them in the following sections.

Planning and requirement analysis

Since the requirements analysis is the first stage, it is the most important and fundamental stage in SDLC. This stage starts by identifying the client's stakeholders, and then conducting several meetings and workshops to define the expectations and gather the requirements.

This stage is performed by the business analyst, the project manager, and the senior technical members of the team. They conduct meetings and workshops with the client to gather all the functional and non-functional requirements, such as the purpose of building the product, what problems it will solve, how it will improve the efficiency of the work, what it will include in terms of functionalities and services, who the target audience or the end user is, identifying the user journeys, detailed use cases and test cases, hardware requirements, backup strategies, and failover processes.

Planning is the process of creating a detailed but high-level plan for how and when each module or task in the project will be developed. The aim is to identify the tasks and their dependencies, along with the expected output of each task. This should be aligned with the client's expectations, as defined in the requirement analysis.

After this stage, everyone in the team should have a clear view of the scope of the project, including its budget, resources, and deadline, as well as possible risks and quality assurance needs. This will be shared with the client to align them with the execution of the project and to give them better transparency.

Let's take a look at the different techniques and activities that we usually use when executing the requirements analysis phases:

- **Use cases**: This is an effective technique that is widely used to capture user requirements. It allows us to identify the possible flow of each feature to be implemented in the system, along with how it will interact with the end users. You may be wondering, *how many use cases should I write?* This might sound difficult at first, but the simple answer to this question is to make sure that you write down as many use cases as possible, to make sure you cover all possible actions and functionalities that should be included in the system.

  The following are the common sections of a use case:

  a. Use Case Name

  b. Summary Description

  c. Actors

  d. Pre-Conditions

  e. Post-Conditions

  f. Level

  g. Stakeholders

- **Business Process Modeling Notation (BPMN)**: This is used globally to create graphs that describe and document a business sequence using symbols and elements. This technique is recommended if you are implementing business automation processes or the product contains business workflows such as approval cycles.

Here are the basic shapes of BPMN diagrams in **Visio**:

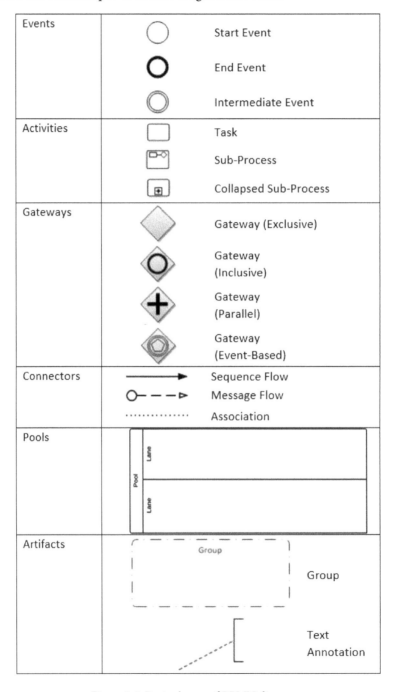

Figure 1.2: Basic shapes of BPMN diagrams

- **Unified Modeling Language (UML)**: UML is used to create flowcharts and diagrams to visualize and document software components, such as classes and interfaces. UML is a good design practice and a very useful technique for creating object-oriented software; it helps software developers model and communicate any complex architectural software design:

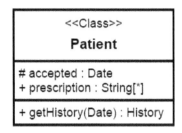

Figure 1.3: Sample class diagram

- **Flowchart technique**: This is another graphical representation that's used to describe the different steps of a sequential and logical process flow. In the following diagram, we have a sample flowchart for a checkout process. The green box is the starting point for when the user attempts to add items to the checkout cart before settling the payment and receiving confirmation at the end. The red box represents the end of the process; that is, its completion:

Figure 1.4: Sample flowchart diagram

- **Data flow diagram (DFD)**: A diagram is worth a thousand words. You can use a DFD to visually represent the way data flows through a process or service in the system. This diagram is used to identify and describe the input data and how it is moving through the system to reach its storage location and form the output data. Here is a sample DFD diagram describing the flow of the data in a purchase order process:

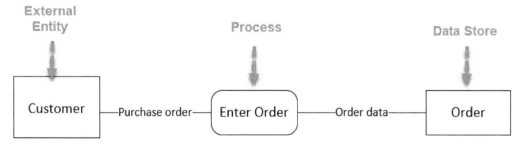

Figure 1.5: Sample data flow diagram

- **Role Activity Diagrams (RAD):** This is a role-oriented representation of every possible action in the system. It is used to easily describe and visualize the different roles that are involved in executing each process or service in the system. The following is a sample role activity diagram describing an ATM transaction and showing the steps that are accomplished by each key role:

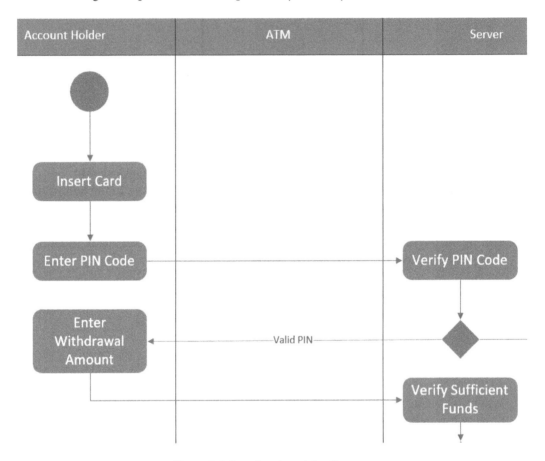

Figure 1.6: Sample role activity diagram

- **Gantt charts:** These are used in project management to assist with planning and scheduling projects of all sizes. They provide a visual representation of tasks, their delivery dates, and the order and dependencies of each task.

  This makes the execution plan more simplified and transparent for the client. The following is a sample Gantt chart representing a project plan. The tasks are grouped based on a specific context and linked through the predecessor column, along with the start date and end date:

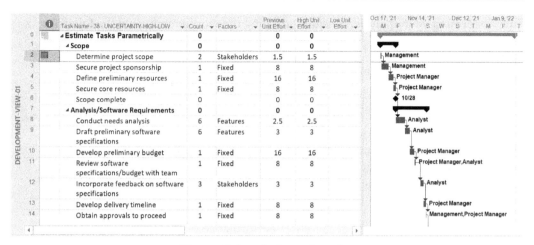

Figure 1.7: Sample Gantt chart plan

- **Gap analysis**: This is a technique that helps compare the current actual results of the system with what was expected by the client in the early stages of the project. It helps denote any missing strategic capability or feature in the system. It should also recommend ways you can make improvements that will help the client meet their initial targets. The following is a sample template that can be used to conduct a gap analysis exercise:

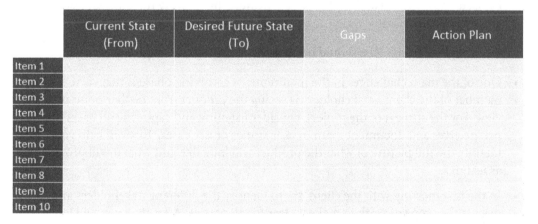

Figure 1.8: Sample gap analysis template

- **Building prototypes**: Building a mock-up, or a **Minimum Viable Product** (**MVP**) model, of the product will give the end users an idea of what the final version of the product will look like once all the features have been implemented. Using this technique, you can identify any feasibility challenges that you may face when you actually implement the product.

While performing your analysis, consider grouping the requirements into the following three categories:

- **Functional requirements**: These represent all the detailed features and functionalities of the system. They are very important for both the development team, to find out what to implement, and the client's stakeholders, to help them align on the final results of the product.

- **Operational requirements**: These define the scenarios and the performance measures, along with the associated requirements, that are needed for the product to operate properly in accordance with the client's expectations. This includes the following:

  a. Establishing critical and desired user performance

  b. Defining constraints

  c. Establishing the infrastructure needed

  d. Establishing measures of effectiveness

- **Technical requirements**: These describe the technical part that must be fulfilled to easily and successfully deploy the product and make it functional with good performance, as per the client's expectations. This includes the technology that will be used, the technical architecture, the hardware, third-party integration, testing, and deployment plans.

Here are a couple of things that should be considered during this stage:

- One of the main challenges in the requirements gathering phase is that each member of the client's stakeholders is seeing the product from his/her point of view. For the success of the project, consider listening and capturing all users' perspectives and document them in user stories or use cases. This will help you identify the full picture of what the product will look like and what it will provide as features.

- In the first meeting with the client, try to identify the different stakeholders and discuss the scope of work to make it clear for all parties. After that, you will have to meet with all the stakeholders to collect the detailed requirements. During these workshops, make sure you keep all your discussions within the scope set forth. This is important to keep the requirements aligned with the business needs and to avoid adding functionalities that the product was never expected to provide.

So far, we've explored the different activities and techniques we can use for planning and conducting the requirement analysis workshop, which is essential for the success of the project. In the next section, we will learn how to document requirements.

Defining requirements

The next step after completing the requirements analysis workshop is to document all the information that was gathered in the previous step to define the product requirements. Usually, the output result of this activity is the **Software Requirement Specification (SRS)** document, which consists of all the detailed requirements to be designed and developed during all the phases of the project, from the beginning to its end, until the desired product is delivered. This SRS becomes the *requirements contract* that will be used to develop the product. It will address all the business needs of your client.

Once the SRS document has been finalized and reviewed by all the parties involved in this project, make sure you send it back to the key stakeholders, or the representatives of the key stakeholders, to sign it. The purpose of signing the SRS is to agree that the requirements that are presented and defined in the document are clear and reflect the business needs, as discussed in the analysis workshop. This formal commitment, which is expressed by all parties involved, will play a crucial role in the project life cycle to ensure that the project will not struggle from **scope creep** during its implementation.

> **Important Note:**
>
> In project management, scope creep (or requirement creep) refers to a situation where the client is continuously requesting changes and adding new features to the product, even after project kickoff. As a result, the project's scope will continue to grow, which will affect the delivery time and the final cost of the product. This should not occur and to prevent it, you must make sure that all the business needs (that is, the scope of the project) are very detailed and properly defined, and that the client has officially committed to the scope of work.

A basic outline for an SRS document may look like this:

1. Introduction

    1.1 Purpose

    1.2 Intended Audience

    1.3 Intended Use

    1.4 Scope

    1.5 Definitions and Acronyms

2.   Overall Description

2.1 User Needs

2.2 Design and Implementation Constraints

2.3 Assumptions and Dependencies

3.   System Features

3.1 Functional Requirements

4.   External Interface Requirements

4.1 User Interfaces

4.2 Software Interfaces

4.3 Hardware Interfaces

5.   Non-Functional Requirements

5.1 Performance Requirements

5.2 Security Requirements

5.3 Software Quality Attributes

Feel free to use this outline and modify it as per your needs, but keep in mind that this document should describe the functionality the product needs to fulfill, along with the technical specifications. Therefore, it should be simple, easy to read, and understand by the project stakeholders. In the next section, we are going to learn about the architecture design phase.

Architectural design

How will we build the product? This a crucial question to answer, especially if you're building a complex or large-scale product that will be used by a wide range of users.

To answer this question, we need to start the **architectural design phase**, which consists of converting the software specifications that were defined and documented in the previous stages into an abstract design specification plan called the **architectural design**.

The starting point of this phase is to go through the SRS document and understand every single detail in the requirements. This will help you create the best architecture design, which will ensure you deliver a high-quality product. It is the responsibility of the technical team to document their design in a **Design Document Specification (DDS)** document. The intended audience of this document is the designers, software developers, and QA testers.

The purpose of this document is to present a comprehensive architectural overview and depict all the technical details of the system components. More specifically, it should present the following:

- The system architecture, components, classes, their attributes, and methods
- The database's design, including the definition of the tables and fields, along with the relationships between tables
- The graphical interface design
- Hardware or software environment
- End user environment
- Security requirements
- Performance requirements and capacity limitations

This DDS is reviewed by all the key technical stakeholders. Based on various factors such as design modularity, performance, security, capacity limitations, risks, budget, and time constraints, the best design approach is selected to build the product.

A basic outline for a DDS document may look like this:

1. Introduction

    1.1 Purpose

    1.2 Scope

    1.3 Design Goals

    　　1.3.1 Maintainability

    　　1.3.2 Optimized Performance

    　　1.3.3 Designed Friendly

2. System Overview

    2.1 Algorithms

    2.2 Technologies Used

    2.3 Architecture Diagrams

    2.4 Database Design

3. Design Considerations

    3.1 Assumptions and Dependencies

    3.2 General Constraints

    3.3 Goals and Guidelines

    3.4 Development Methods

4. Architectural Strategies

    4.1 Strategy-1 name or description

    4.2 Strategy-2 name or description

    4.3 ...

5. System Architecture

    5.1 Component-1 name or description

    5.2 Component-2 name or description

    5.3 ...

6. Policies and Tactics

    6.1 Policy/tactic-1 name or description

    6.2 Policy/tactic-2 name or description

    6.3 ...

7. Detailed System Design

    7.1 Module-1 name or description

    7.2 Module-2 name or description

    7.3 ...

8. Traceability

9. Glossary

10. Appendix

You can use this outline to describe your architecture and prepare the DDS document. The more you make it clear and detailed, the more you make it easy for the developers and testers during the implementation and testing phases. Next, we will explore the development phase.

Software development

In this stage of SDLC, the software developers start actually developing the product. The technology that's used and the programming language, including all the technical standards, should be aligned with what was agreed on in the DDS document. Keep in mind that the development activities can be accomplished very smoothly when the design specifications are detailed and organized in a proper manner.

Testing

Did we get what we want? Testing the product is a must before launching it to the end users. This stage starts alongside the development stage, where the developers are responsible for testing what they are developing. At this time, it is just basic testing and not enough to say that the product is ready to go live.

Therefore, an official testing cycle should be conducted once the development activities of a specific module or the entire set of features have been completed. During this phase, several types of testing should be conducted, every single functionality should be tested thoroughly, and the identified defects should be reported to the developers to get them fixed.

The quality assurance team can use the test cases that have been documented in the SRS, or they can refer to the use cases to test the product. It is recommended to run the test cases every time the developers release a new version of the product until it reaches a stable version. This is to make sure all the defects that were reported in the previous cycles have been closed.

Deployment and maintenance

Software developers tend to invest the majority of their time in the design and development activities of the product, which is good. Despite its importance, I have learned from several projects that this is not enough. Setting a strategic plan for deployment and maintenance is a key factor for the success of the product.

The focus at this stage is to make the product available for end users so that they can start using it. To do so, the product should be deployed to the production environment.

First, it is recommended that you deploy the product in a testing or staging environment. This is where the **User Acceptance Testing** (**UAT**) activities should be performed. All the issues will be solved and deployed back to this environment. Once the product reaches a stable version that is accepted by the client and meets all the specifications that were approved in the previous phases, the product can be moved to the production environment.

> **Important Note:**
>
> User acceptance testing is the final round of testing. It is performed by the client to verify that every single functionality provided by the software works, and to confirm that all the requirements have been covered. This will ensure that the software behaves exactly as the users expect and that they can easily use it without any errors or crashes occurring. At the end of the UAT, the client should accept the software or request some improvements before moving the software to the production environment.

The maintenance phase starts immediately after the product is fully operational in the production environment and signed off by the client. This is a crucial step from the client's point of view because it ensures that their product continues to perform as designed after its deployment.

Types of maintenance

There are four types of software maintenance:

- **Corrective maintenance**: This is used mainly to rectify some errors and faults that are observed while the system is in use or to improve the performance of the system.

- **Adaptive maintenance**: This may be needed when the client requests to run the software on a new environment such as new hardware or a new operating system. Sometimes, clients request to move their products from an on-premises environment to **Azure Cloud**. Moreover, it can cover integrating the product with third-party software.

- **Perfective maintenance**: This type of maintenance focuses on implementing new features in the product. These features can be requested by the client to accommodate new business cases, or they can be reported by users who have already started interacting with the product and noticed some missing functionalities that can help facilitate their work and improve the overall experience.

- **Preventive maintenance**: This is commonly used to detect and correct errors that may cause software failure in the future. It helps reduce the risk of the issues that aren't significant at this moment but may cause serious problems in the future; for example, assuming the clients are expecting to have an increased number of users who will start using their product after 2 months, but this load cannot be accommodated by the current environment's specs. In this case, planning and updating the software environment in advance to serve the load that will be caused by the new users is considered preventive maintenance.

Let's take a look at the following table to understand when and why we should apply these maintenance types:

Software Maintenance			
Corrective	Adaptive	Perfective	Preventive
You need it when the released software doesn't work as expected or as per the requirements	You need it when changing the operating environment where the software was initially installed	You need it when the client identifies a better way of doing a specific functionality	You need it when identifying a potential issue that may occur in the future while using the software

Figure 1.9: Software maintenance types

In the next section, we are going to explore the difference between software maintenance and warranty.

Maintenance versus warranty

People may get confused about maintenance and warranty. A software warranty is a formal and legal guarantee that the product will perform properly, as per the specifications, for a certain period. It is a promise to fix any errors or malfunctions in the system at no cost during the warranty period.

The maintenance agreement is sold to the client for long-term and ongoing maintenance activities such as upgrades, updates, or product enhancements.

We have just explained the different stages of the SDLC and highlighted the expected output of each stage. In the next section, we are going to provide an overview of the popular SDLC models.

In this section, we explored all the SDLC phases, from planning and requirements analysis to deploying and sign-off. In the next section, we will get to know the most popular SDLC models.

Getting familiar with the popular SDLC models

Every product requires a suitable approach to developing it. Usually, this decision is made based on multiple factors, such as if the requirements are well-documented, the requirements are not ambiguous, the project is short, and so on. In this section, we will highlight some of the most popular models that are used in software development.

The Waterfall model

The **Waterfall model** is a straightforward and sequential approach to building a software product. Each stage of the development cycle should be completed before you move on to the next stage and usually, the output of each stage is considered to be the input for the next stage.

Here is a representation of the different stages of this model:

Figure 1.10: Waterfall stages

Some of the advantages of the Waterfall model are as follows:

- Stages are clearly defined and easy to understand

- Stages are well-documented

- Works well for smaller projects where the requirements are well-defined

Some of the disadvantages of this model are as follows:

- The working version of the product will be delivered at a late stage of the development cycle.

- Not a good model for complex and ongoing projects since the stakeholders won't be able to give their feedback at the early stages of the development process.

- Not a good model when there is a high risk of requirements changing.

The Agile model

The **Agile model**, an example of which is Scrum, is one of the most well-known development methodologies and is widely adopted by many IT organizations. It is also applied to non-tech projects.

The approach of this model is to break the product into cycles or iterations. Each iteration lasts for about 2-4 weeks (usually, it shouldn't be a long time). At each iteration, the development team should deliver a complete working version of the software. The idea is to take the use cases and split them into iterations so that you get a functioning part of the product at the end of the iteration. In this way, the development team is producing ongoing and incremental releases that have been well tested.

This approach helps teams identify and address issues early on. It also involves the stakeholders throughout the development process to get their feedback.

The following diagram is a quick representation of the Agile stages:

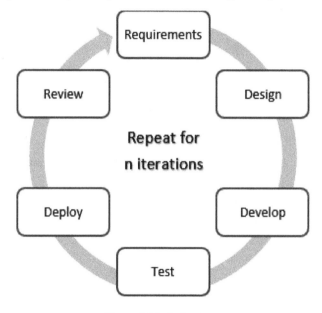

Figure 1.11: Agile stages

The Spiral model

The **Spiral model** is a combination of the Iterative model and the Waterfall sequential model. Usually used for large projects, it provides support for risk handling at the early stages of each iteration. With this model, the project passes through four phases:

- Identifying objectives by gathering the business requirements
- Performing risk analysis
- Reviewing and evaluating
- Developing and testing

Here is a diagram depicting the Spiral model:

Figure 1.12: Spiral model

With each iteration, you can build a prototype of the new feature and functionalities that will be delivered in this iteration.

These phases are repeated in a *spiral* until the entire product is delivered, allowing for multiple rounds of refinement.

The advantages of the Spiral model are as follows:

- This model provides an early indication of the existence of risks.
- Critical high-risk functionalities are developed first.
- Stakeholders are closely tied to the entire development life cycle phases.
- Users can see the system in action at early stages with the use of prototypes.
- Stakeholders can incorporate early and continuous feedback.

The disadvantages of the Spiral model are as follows:

- This model is costly and is not recommended for small projects that have low risks in most cases.

- Managing the process is somewhat complex.

- Risk assessment expertise is required to run this model.

The DevOps model

In a **DevOps model**, the developers and operations teams work together. You may be wondering, *well, what does this mean?*

Using the traditional models that we talked about earlier, companies were splitting up their resources into teams that handled specific responsibilities:

- A development team to architect and build the product.

- An operations team to prepare the environment and host the product.

- A test team to prepare the test cases and conduct thorough QA testing and to report back to the development team.

With the DevOps methodology, the developers and operations teams are requested to collaborate closely – as one team – in all the stages of the SDLC process. A successful DevOps model ensures continuous feedback, accelerates the deployment, improves the development process, and automates manual processes.

Here is a representation showing the different steps in the DevOps model:

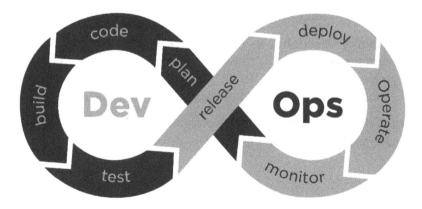

Figure 1.13: DevOps model

The advantages of the DevOps model are as follows:

- Fast delivery of features
- Better responsiveness to problems
- Efficient operations
- Reduced bottlenecks
- Better communication and collaboration
- More productive team members, with more time to innovate

The disadvantages of the DevOps model are as follows:

- DevOps requires culture change and new methods of communication, which is a big challenge in a traditional environment.
- There is a need to upgrade the infrastructure to optimize the process, which can be expensive for some companies.
- Fast development can lead to critical security shortfalls.

Now, let's learn how to choose the right model.

Choosing the right model

When selecting the right SDLC model to build a particular product, it's important to remember that each model offers a unique process that may help you overcome the challenges that you will encounter during the development cycle. One model would never fit every project or every client's needs, which is why you should understand these popular models and know when to apply them.

Finding the right model depends heavily on the factors the project will be executed with, such as your current infrastructure, the culture adopted by your team, and how the client would like the project to be managed. Certain projects may run best with a Waterfall approach, while others would benefit from the flexibility of the Agile model.

Let's take a look at the following table, which highlights the main factors when it comes to choosing the right model for your product:

Model name	When to apply it?
Waterfall model	You can apply it when the project has very clear and documented requirements and when the project is initiated from a **request for proposal (RFP)**.
Agile model	You can apply it when the customer is requesting to have some functional requirements ready in a short period of time. Remember that this model requires engagement and close interaction with the customer. This model should allow you to have a minimum viable product in a short period which also increase the customer satisfaction.
Spiral model	You can apply it in large applications which are built in small phases or when releases are required to be frequent and when you need to evaluate risks and costs for each phase.
DevOps model	You can apply it if you want to automate the entire development cycle. This requires changing how the development and operations are done. DevOps can be applied under the following conditions: • If you want to apply an agile development process • If you want to adopt cloud infrastructure to manage the entire development cycle • If you want to automate your software testing and deployment

Figure 1.14: How to choose the right SDLC model

In this section, you explored the most popular SDLC models. Each one offers a unique methodology that can help you overcome different challenges you may encounter in your career. You also learned how to choose the right model for your product.

Summary

In this chapter, you learned about the definition and the importance of the SDLC, as well as how it can help the organization deliver products in an efficient way. Then, you learned about the different stages of the SDLC, the most popular models, along with their advantages and disadvantages, and how to choose the right model for your team.

In the next chapter, you will learn about the different team roles, along with their responsibilities, and how they fit into the SDLC process and the team structure.

2
Team Roles and Responsibilities

In the previous chapter, we covered all the essential phases of the software development life cycle. Now, let's get to know the different team roles contributing to the execution of these different phases.

Employees are the most important assets of an organization. One of the key factors of a successful software project is ensuring that the key members of the development team are all in place. The success of a project also depends on how well the team collaborates and communicates efficiently in order to deliver the best outcome. This chapter focuses on the main roles within a typical software development team and their corresponding responsibilities.

In this chapter, we will cover the following topics:

- Exploring the development team hierarchy
- Highlighting the five key attributes to consider when assembling a team

By the end of this chapter, you will have learned about the typical hierarchy within a software development team and what to expect in terms of the responsibilities of each member.

Exploring the development team hierarchy

In many situations, clients wonder why we allocate different roles and specialties to build a product. Indeed, they expect that a software engineer role is enough. This chapter is intended to answer this concern and to highlight the specific role that every member plays in an Agile development team in order to deliver the best possible performance.

Typically, when you start assembling a software development team, the decision to choose the roles, along with the responsibilities of each member, depends on the answers to these two questions:

- *What type of product will you develop?*

- *What is the methodology of work that will be used?*

The following diagram shows the key positions of an Agile software development team. You will notice that we have highlighted the solution architect position; they play a liaison role between the technical and non-technical teams. This is the person who will design the architecture of the solution:

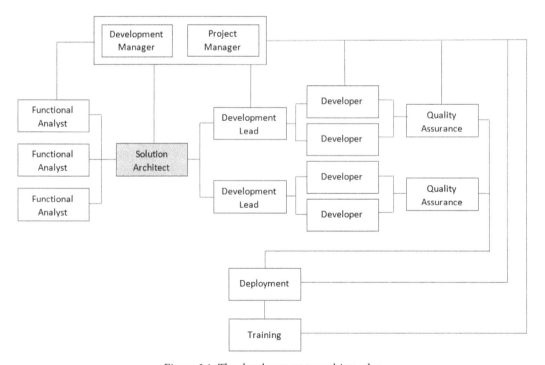

Figure 2.1: The development team hierarchy

Let's dive deeper into each role and its responsibilities within a team.

The project manager – the godfather

The **project manager** (**PM**) is an organized and detail-oriented individual with good knowledge of project estimation techniques. The PM is responsible for knowing the key stakeholders of the project and effectively communicating with each of them to plan, schedule, prepare the budget, execute tasks, and ensure the delivery and completion of the software product.

The duties and responsibilities of the PM can include the following:

- Planning the entire phases of the project from analyzing requirements to testing and maintenance

- Deciding on the methodology that will be used for the project in coordination with the client

- Allocating all the resources that are needed for the successful completion of the project and ensuring they have the right environment in terms of software tools and hardware to implement and test the project

- Proposing the project timeline and scheduling the tasks of each phase

- Leading and supervising the successful execution of each phase of the project by assigning tasks to team members and ensuring they are delivered on time

- Motivating team members to deliver good quality output

- Creating the project budget and providing regular status reports to senior management

- Ensuring the requirements are fully understood by the team members, making and communicating change requests, and ensuring the alignment of the output with the client's expectations

Next, as per the team hierarchy, we will examine the functional analyst role.

The functional analyst – the explorer

The **functional analyst** is responsible for ensuring that all requirements have been thoroughly discussed and analyzed with the client. After doing so, the requirements are captured, documented correctly, and communicated clearly to the team before the kick-off of any development activity.

They play an important role in translating the business processes, no matter how complex they are, into logical and functional requirements that can be developed by the tech team. Their role is to bridge the gap between the business users and their expectations with the development team, which is responsible for building the product.

This role might have different titles, such as requirements analyst, system analyst, or business analyst, but they all, more or less, deal with the same responsibilities.

Some of the duties of the functional analyst include the following:

- Meeting with the client stakeholders, managerial team, and business users and gathering requirements

- Identifying the primary goals of the product to build

- Analyzing technical and business requirements and documenting them

- Ensuring that the requirements are properly communicated and explained to the development team

- Testing the final product to validate the objectives and guarantee the compliance of the outcome with the business goals of the client and their users

Next, we will explore the responsibilities of the solution architect.

The solution architect – the game changer

A **solution architect** is responsible for leading the technical design and managing the overall engineering side of the solution concerning specific business requirements.

This team member should have a balanced combination of technical and business skillsets in order to create the solution architecture.

A solution architecture can be a multipart process with a wide range of issues that are focused on particular audiences and business objectives. Therefore, the main focus is to analyze and understand all parts of the business model, including all of the requirements defined in the early stages.

Then, they need to design a specific solution and introduce the overall technical vision for the solution that should fit the current environment along with the client's expectations. The solution architecture will be communicated to the rest of the development team, who will then use the design specifications, in addition to the requirements, to implement the solution.

Some of the duties of the solution architect include the following:

- Analyzing and understanding the requirements and then proposing an architectural design of the solution

- Assessing the current technologies and systems in place to introduce integration links between the product to be built and the existing systems to meet the client's needs

- Being transparent to all stakeholders and informing them about any technical issues with the current product being implemented and then proposing solutions

- Taking technical decisions after assessing the business impact they might have on the final product

- Documenting proposed solutions and monitoring all requested updates to make sure that they have no impact on the overall design of the solution

- Recommending the right hardware for the product to function properly and coordinating with the IT professionals to prepare the necessary environment to build, test, and host the product

- Tackling common project challenges, such as team skills, communication conflicts, unclear requirements, and unrealistic deadlines

- Identifying possible risks in advance to prevent any surprises during the implementation

- Coordinating with the project manager and team leader to prepare the project timeline and detailed scheduling, which will be the main input for pricing the work needed

- Coordinating with the developers and team leaders to resolve technical problems as they arise

- Guiding the development team by regularly researching existing technologies and proposing changes and new techniques to the development team to improve processes

This role requires deep technical knowledge and hands-on experience in the following areas:

- Business analysis to help understand and improve the business processes by translating them into functional use cases

- IT infrastructure to recommend the right environment for the product

- Software architecture design to propose a modern solution architecture for the product

- Cloud development to develop and deploy solutions to the cloud, as well as streamline the **SDLC** when shifting to the cloud

- **DevOps** to help to improve the Agile development life cycle

Next, we will get to know the responsibilities of the development lead.

The development lead – the tech-savvy one

The **development lead** is responsible for leading and supporting the developers to implement all the technical aspects of the product. However, before starting the development, they work closely to provide an accurate estimation for the work that is needed. This assessment is used by the project manager to create the project timeline and provide a structured breakdown of all tasks.

The technical lead should effectively communicate with the project manager to provide regular progress reports of the development activities. On the other hand, the technical lead should also communicate with the solution architect so that technical issues, changes, or conflicts can be tackled at the right time and in a professional manner.

One of the main responsibilities of the technical lead is to enforce the coding standards and best practices with the team members and to ensure that they are clear and easy to apply so that the developed code can be reusable, readable, and extendable at any point in time.

Some of the duties of the development lead include the following:

- Managing day-to-day assignments and organizing team initiatives
- Conducting technical training sessions to coach team members on new technologies and techniques and ensuring they all apply the same standards
- Evaluating the performance of the team and suggesting improvements and goals
- Continuously listening to team members' feedback and concerns to resolve any issues or conflicts that might affect the team's spirit or the progress of work
- Motivating team members to improve their analytical thinking and creativity
- Suggesting and organizing team-building activities
- Developing team strengths and improving weaknesses
- Recognizing team achievements and coordinating with the management to reward accomplishments

Next, we will explore the career tracks of software developers.

Software developers – the masters of magic

Software developers are responsible for properly understanding the requirements defined in the early stages of the project. Then, they begin developing the modules and functionalities as per the schedule that has been agreed with the project manager and the client. Any ambiguity in the requirements that can't be clarified by the team lead should be discussed with the functional analyst or the solution architect. This is very important in order to reduce project risks and to ensure good quality deliverables, which will lead to the success of the project.

There are three career tracks for a software developer to follow:

- **Frontend**: In this case, they are responsible for delivering the client-side blocks of the product. Usually, they use frontend programming languages, such as **HTML** and **JavaScript**. They should be skilled in **jQuery, CSS (SASS** or **LESS)**, and responsive frameworks, such as Bootstrap. In modern web development, frontend developers are very popular with the existence of JavaScript frameworks, such as Angular, React, and Vue, which are used to build single-page web applications:

    a. **Angular**: This is an open-source JavaScript framework developed by the Angular team at **Google**.

    b. **React**: This is an open-source JavaScript framework developed by the React team at **Facebook**.

    c. **Vue**: This is an open source MVVM JavaScript framework created by Evan You, who formerly worked for Google on the **AngularJS** framework.

- **Backend**: In this case, they are responsible for developing the server-side functionalities and blocks of the product, such as the web services or the web API. They are also responsible for designing and creating the databases, including all SQL queries and transactions. Some of the main programming languages and skills that every backend developer should have include the following:

    a. .NET C# (.NET Framework and .NET Core)

    b. ASP.NET MVC and Web API

    c. N-tier architecture

    d. Entity Framework and ADO.NET

    e. MSSQL databases and queries

    f. Azure Blob storage, Azure App Service, and Azure Functions

    g. Microsoft Power Automate

h. Deploying Docker containers on Azure

i. Knowledge of HTML, JavaScript, and jQuery

j. A JS Framework such as React, Angular, or Vue

k. Unit testing

l. Team Foundation Server or Git for source code control and versioning

m. Azure DevOps to manage the source code or the development cycle

- **Full-Stack**: This role is a combination of the previous two tracks.

Some of the duties and responsibilities of the software developer include the following:

- Researching, designing, and implementing clean and efficient code based on the requirements, specifications, and coding standards

- Coordinating with the team lead on assignments and the day-to-day schedule

- Troubleshooting the code to fix reported issues with the ability to propose workarounds

- Deploying and testing the product in the test environment as well as in the production environment

- Providing professional code documentation and supporting the preparation of the user guide documentation

- Contributing to the code review sessions to improve quality and enhance overall performance.

- Providing status reports to the team lead, as needed, on a daily or weekly basis.

- Coordinating with the team lead to solve any blocking issues that need advanced expertise to be resolved.

Next, we will learn about the responsibilities of QA engineers.

Quality assurance – the quality guards

QA engineers or **software testers** are responsible for conducting full cycle testing for a product to ensure that all requirements and use cases have been developed and that the product is free of defects. To achieve this target, they should create test cases out of the requirements or the use cases. The goal is to test every single function or process in the product and then coordinate with the developers to provide steps in which to reproduce the defects. Once the defects are fixed by the developers, the QA team should conduct another cycle of testing to confirm the fixes.

Some of the duties of QA engineers include the following:

- Coordinating with IT professionals to ensure that the testing environment is ready before starting the testing of the product

- Creating test cases and running test plans to test all the functionalities of the product

- Identifying the defects and logging all the steps and details to help the developers resolve these defects

- Monitoring performance and generating metrics, which will help to improve the efficiency of the product

- Conducting security testing to identify security threats and prevent vulnerabilities

So far, we have discussed the various roles, along with their specialties, of a typical Agile team. In the next section, we will highlight certain personality traits and additional things to consider when assembling a team to build your product.

Highlighting the five key attributes to consider when assembling a team

Building an effective and goal-oriented team with a clear purpose can be challenging because it brings together different cultures, attitudes, and communication skills. If you talk to senior managers who are responsible for managing teams, they will tell you that the most significant problems they face are related to the communication issues and internal processes adopted by the team to do the work that is needed. That's why it is important to have, within your processes, a kind of protocol, team norms, standards, or best practices that can assist the team to get along and build effective interpersonal relationships to accomplish their goals.

In the following subsections, we'll walk through some key attributes that need to be considered when you're assembling a software development team.

Building a great team culture

Hiring team members based on their technical qualifications is important. However, you don't want to hire a team that has no harmony and can't communicate as one entity to properly achieve your goals.

Build a strong team by choosing members who have chemistry, can work together, have the same vision, and can collaborate effectively. Support your company culture by considering the following personal traits when evaluating candidates:

- Has a positive attitude
- Is a team player
- Is self-motivated
- Has a strong work ethic
- Is detail-oriented
- Is a good communicator
- Is a self-improver
- Is adaptable
- Is honest

Establishing development standards and best practices

Having rules and standards can lead to good results and better code maintainability. There are a lot of common standards out there that you can apply. However, it is important to keep the standards relevant, simple, and clear to everyone on the team.

As for the best practices, they are a set of methods or techniques that have either shown a good outcome or an improvement to the process. Therefore, it is recommended that you analyze the current processes in place and suggest best practices based on previous experiences.

Equipping the team with the right tools

They say *you're only as good as the tools you use*, so it is vital that you provide your team with tools that can help to improve team collaboration and allow your team to perform tasks more effectively, which will reduce the turnover.

Here is a list of must-have and recommended tools that will support you in the building of your .NET product:

- **Visual Studio and Visual Studio Code**: This is the primary development IDE for .NET solutions.

- **Azure DevOps**: This is an online platform provided by **Microsoft** that offers DevOps tools to develop, test, and deploy software products.

- **Azure Storage Explorer**: This is a free tool from Microsoft that allows users to easily browse and manage multiple Azure Cloud storage accounts.

- **Service Bus Explorer**: This is a tool that allows users to easily connect to a Service Bus namespace and execute management and data operations.

- **Notepad++**: This is a free tool used as a text and source code editor. It supports several languages, and it is a great replacement for the regular **Notepad**.

- **Postman**: This is a collaboration tool that is used for API testing.

- **Snagit**: This is a screenshot and video capture tool (you can use the **Windows Snipping Tool** if you can't afford Snagit's license).

- **GitHub Desktop**: This is an open-source tool that simplifies your development processes.

- **PowerShell**: This is a command-line shell from Microsoft, which is used to automate tasks by writing scripts in **PowerShell** and scheduling their execution triggers using Windows Task Scheduler.

- **Fiddler**: This is a debugging tool that is used to inspect HTTP and HTTPS requests between your development server and the web server.

- **NuGet Package Explorer**: This is an open-source tool with an easy-to-use GUI that allows you to create and explore **NuGet** packages.

- **Regex101**: This is a tool that allows you to generate and test regex syntax.

- **JSFiddle**: This is an online IDE tool that you can use to test and showcase HTML, CSS, and JavaScript code snippets.

- **U2U CAML Query Builder**: This tool allows you to easily construct **SharePoint CAML** queries.

Next, let's learn how to improve the communication within your team effectively.

Maintaining continuous communication

One of the key factors regarding the running of a successful team is effective communication. Here are few tips to improve team communication:

- Encourage two-way feedback and practice active listening.
- Team members should have clarity about their roles and expected responsibilities.
- Build team spirit by introducing team activities.
- Make use of cloud tools to collaborate, and decrease the number of follow-up meetings that can sometimes cause distractions. This will allow you to use your time wisely.
- Offer training sessions to help develop necessary communication skills.
- Encourage the acceptance of constructive criticism.
- Continuously evaluate team communication and suggest improvements.

Helping developers grow professionally

Employee retention is very important; companies should take initiatives to increase employee satisfaction and keep them committed to the delivery of good quality work and to drive more productivity.

One way to pursue this is by setting a professional growth plan for each member of the team based on their roles and responsibilities.

Here are few tips to help your team members grow professionally:

- Identify the soft skills needed for each team member and suggest actions they can take to improve.
- Always give recognition and rewards.
- Regularly evaluate team members, not just during annual reviews.
- Set goals and plan training programs for each team member.

In this section, we highlighted the key attributes that could affect the effectiveness of your team.

Summary

Your role as a solution architect is to build and lead a team to deliver successful projects; this can't be achieved if you don't have good team-building skills. This chapter is intended to help you unite a successful development team, which is composed of different roles that participate in the SDLC process.

In this chapter, you learned about the different roles identified in a typical software development team, their responsibilities, and how they interact with each other to get the job done. Next, we highlighted some key attributes to consider when assembling a team, such as team culture, applying standards and best practices, strong team communication, and professional team growth.

In the next chapter, you will look at a quick overview of what solution architecture actually is. Then, you will learn about the role of the solution architect and their related responsibilities. After that, we will highlight the key personality traits that will support you in becoming an effective solution architect.

3
What Makes an Effective Solution Architect?

In the previous chapter, we highlighted the importance of the different roles within a software development team. We also explored some key attributes to support you in assembling a powerful and successful team. *Why do you need to learn this?* Because you might hire the best candidates out there, but if you don't have a proper culture in place, and if the team members are not able to communicate with each other efficiently, you could end up losing the project or your client.

In this chapter, we will primarily focus on the solution architect role. Additionally, we will elaborate on a set of personality traits that should empower you to lead effectively in today's digital world.

In this chapter, we will cover the following topics:

- An overview of solution architecture
- Exploring the fundamental soft skills that every solution architect should have
- Getting to know some common pitfalls that should be avoided
- Learning the difference between an enterprise architect, a technical architect, and a solution architect

By the end of this chapter, you will have an overview of what solution architecture is. Additionally, you will get to know core personality traits that you should acquire as a solution architect to empower your architectural thinking and leadership skills.

These personality traits are essential in order to become successful because the solution architect role has to deal with many aspects that are not technical, such as team building, negotiating with clients, resolving conflicts, improving business processes, and creating a culture of innovation and professionalism within the team.

What is solution architecture?

Before we dig deep into the personality traits of the solution architect, first, let's get to know what solution architecture is. **Solution architecture** is a set of activities that aims to explore and analyze a business problem based on predefined requirements and create an architecture design for the proposed technology solution that fits with the client's goals and needs.

Typically, the solution architect should consider the following four key factors and related constraints when creating a balanced and effective solution and its architecture design:

- **Enterprise constraints**: Identify the enterprise constraints and goals behind building the product.
- **Stakeholders' perspective**: Understand and analyze the requirements collected from the business stakeholders and power users.
- **Technology value**: Identify the value of the technology stack and components used in your solution that should comply with the enterprise strategic guidelines and best practices.
- **Project constraints**: The solution architect should consider the project timeline and budget.

Here is an illustration that summarizes the different constraints that should be considered in your solution design:

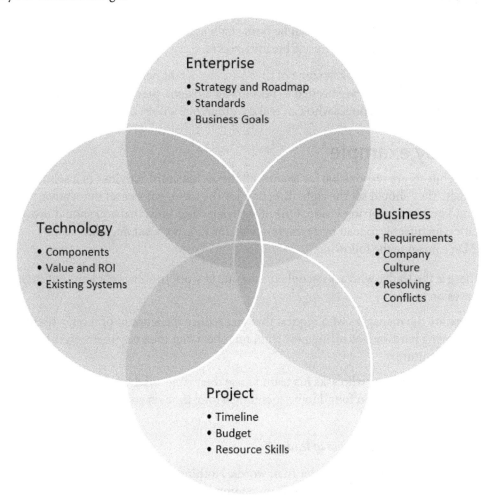

Figure 3.1: An ideal solution should consider these constraints

So far, it looks pretty simple, right? Solution architecture is the process of designing and managing the whole solution engineering, which is expected to solve a business problem by executing a list of practices. This should be accomplished by the solution architect before starting any development activity, and we will examine this in more detail later in this book.

Now, let's dive into some of the essential personality traits that a solution architect should have.

Exploring the personality traits and skills of an effective architect

Being an effective solution architect can be both challenging and game-changing in today's rapidly changing and disruptive business world.

In this section, we'll take a look at some of the most important personality traits, soft skills, and qualities that are needed to become an efficient, practical, and business value-focused solution architect who can make a difference within a company.

Leading by example

Solution architects are responsible for *leading* all the architecture activities of a solution or a product. They should set the right direction for the team, enforce an innovative culture, and build an inspiring vision. One of the leadership skills that you should prove as a solution architect is to lead by example—*walk the talk. So, what does that mean exactly?* Let's imagine the following:

- There is the leader who asks people to stay late at work to solve problems but then leaves on time.

- There are the managers who request that you reduce the amount of money spent on improving hardware or hiring new team members, but then buy themselves luxury office furniture.

- There's a supervisor who asks his team to use their time wisely to be more productive but is then found to be spending a long time on social media during working hours.

Have you ever heard of these types of leaders?

For great leaders, actions speak louder than words. Nothing will kill the enthusiasm or motivation of a team faster than watching an incompetent leader say one thing and then do the opposite. It can be very disappointing, and it leads to low morale, which can be destructive for a team.

As a solution architect, you must know that you have a responsibility toward your team. They are closely watching every move you make because they are looking for guidance. You have to inspire them and push them to get the best outcome. The proper way to do this is by being a good example. Your actions must be consistent with what you say.

Here are some ways that you can use to lead by example and win the trust of your team:

- Get your hands dirty and take responsibility by supporting your team during difficult times. Don't just sit back and tell them what to do.

- Always apply the rules and standards that you set so that your team can follow you.

- Empathy is essential; be emotionally sensitive to the feelings of your team members.

- Take the time to make each individual feel special and important in front of the rest of the team and the entire organization.

- Treat your team members the way you would like to be treated.

- Not all team members are similar; challenge yourself to know them better, accept them for who they are, and respect their unique differences.

- Listen to criticism because, sometimes, it can be constructive. Avoid being defensive when you do so.

- Interpersonal conflicts in the workplace can happen; you cannot avoid them. However, make sure you resolve them as quickly as possible.

- Never take people for granted.

When you walk the talk, you lead by showing your team members how to do things the right way, that is, you set a good example for them. In this way, you become a more effective leader.

Displaying outstanding communication skills

One of the key competencies of a solution architect is having good communication skills, which are essential for building relationships. Remember that your role obliges you to negotiate with clients and resolve any conflicts with the team members. You must be a good listener, not only to respond but to also properly understand the needs of all parties. A lack of communication skills can create a serious bottleneck. However, on the contrary, being able to communicate effectively is a key factor in the success of a project and, therefore, the success of the solution architect.

Here are some tips that you can use in your daily interactions with your team members to achieve effective communication:

- Show empathy because it creates mutual understanding and trust.

- Give compliments to your team members, particularly during difficult discussions. Statements such as *I think what you are saying is great, I agree with you*, and *you did great work so far* will boost their motivation and improve the morale of the team.

- Not all people will have the same opinion as you; be willing to respect the other person's opinion, and don't be rude or arrogant.

- Eye contact is important because it improves the quality of communication and most people consider it as a sign of trustworthiness. Try to look the other person in the eyes while having conversations.

- Respect each other's turn to speak and try your best to not interrupt.

In addition to these recommendations, try to have a clear direction in terms of how you want to manage a particular conversation with a team member or a client. Try to keep it direct; otherwise, it could end up being a useless argument.

The following diagram shows four steps that you can implement to achieve an effective and productive discussion:

Figure 3.2: The steps to achieve a productive meeting

Why is communication important in the workplace? Let's find out:

- Clear communication demonstrates your leadership abilities by describing your goals to the team. Additionally, it allows your workplace to become more collaborative and agile.

- Effective communication has a significant impact on the productivity of the team because it keeps them engaged in important technical or non-technical decisions.

- It creates a positive work environment and improves the relationships that you have with clients and co-workers.

- It boosts the productivity of the team and, therefore, increases the profit of the organization.

Communication skills are vital for conveying your ideas and vision to your team and clients. Remember that the greatest communication skill is listening to others, as that will help you to understand the situation and make proper decisions.

Possessing deep analytical skills

The term *analytical* has become a buzzword in every senior tech job position. *So, what are analytical skills, and why are they essential for you as a solution architect?*

Analytical skills refer to how you investigate a particular problem or business process, collect and analyze all of the related information logically and thoughtfully by understanding how the different elements are connected, research the possible solutions, and then come up with an ideal solution for the situation.

Designing a solution is influenced by different factors; it requires a detail-oriented individual who possesses deep analytical skills with the ability to evaluate various aspects and deal with tasks that require analysis.

These are the skills that you need to find solutions to various problems and difficulties or to help your team members troubleshoot a problem they are facing by proposing the proper solution.

You can develop your analytical skills by getting out of your comfort zone and starting to solve complex problems. Here are some of the core analytical skills that should be mastered by a solution architect:

- **Data analysis**: You must have the ability to analyze the data received, and identify patterns and trends that will support you in your decisions.

- **Communication**: You must be a good communicator to explain your findings to your team, so the client can then describe your recommendations.

- **Critical Thinking**: You must have the ability to analyze complex problems and evaluate the information you have collected to form a rational decision.

- **Creativity**: You should have the ability to go beyond the obvious solution in order to find the optimal one.

- **Research**: You must learn more about the problem you are trying to solve. You can do this by researching online articles and posts that are relevant and learning how other architects or competitors solved a particular problem. It might support you in brainstorming a possible solution.

As a solution architect, you should possess solid analytical skills because they will help you to solve complex problems that might appear during the design or development of the solution.

Showcasing brilliant project and resource management skills

Solution architects are not directly responsible for these two aspects. However, they are expected to focus on business results. Therefore, they are responsible for completing the project in the most efficient way by ensuring the following:

- The business goals of the client are being achieved.

- The project is being implemented within the given timeframes and budget.

- The team skills and assets are properly allocated and are being used efficiently to complete the project.

You are expected to intervene in the five major project management activities:

- **Project initiation**: This is after clarifying the project goals and scope.

- **Requirements gathering and planning**: This consists of developing a work breakdown structure.

- **Project execution**: This is based on the produced schedule.

- **Performance monitoring**: This also includes managing change requests.

- **Project closeout**.

Here is a graphic representation showing these major activities of project management:

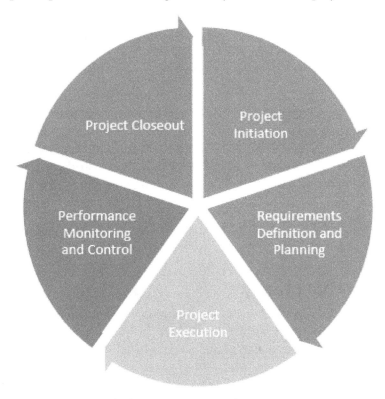

Figure 3.3: The five major activities of project management

The solution architect plays an important role in resource management, which is a critical part of project management. It is the process of planning and assigning resources efficiently to complete the project within the estimated time, budget, and scope, as defined at the earliest stage of the project. In software development, there are two types of resources and assets:

- **Intangible assets** include a variety of non-physical assets, such as staff skills, experience, company reputation, and time.

- **Tangible assets** are physical assets, such as equipment, materials, and investments.

Having project and resource management skills is important for providing a strategic direction throughout the development process by making the right decisions whether they are technical or non-technical.

Exhibiting patience with others

Patience is a human strength. It is considered a tough skill to master; however, if you want to become an effective solution architect, you must have patience, particularly during a crisis. Being patient with others in the workplace is the ability to remain tranquil in the face of difficulties and react positively to your co-workers and clients.

In our fast-moving world, tasks are expected to be completed instantaneously, and delays can create a stressful situation. For instance, your clients will want their projects to be delivered on time or your manager will be expecting a report that you assigned to someone in your team who didn't deliver on time. Think about clients who are continuously changing the requirements, and think about the co-workers who may have bad habits, who are hard to deal with, or who are hard to understand. All of these difficult circumstances are beyond your control, and they can make you instantly frustrated.

Losing patience with these people can complicate the situation. It could damage your relationships and leave a bad image of you.

Working collaboratively

Building an architecture design and delivering a solution is a team effort. An effective architect can work collaboratively with all co-workers (including those who have different skillsets and expertise) with one goal in mind, which is to deliver the project efficiently and meet the client's targets.

Let's take a look at the following reasons as to why collaboration is important:

- It helps you and the team to solve problems faster.
- It boosts motivation and brings people closer together, which can improve retention rates.
- It creates a learning and development environment, and, as a result, team members can learn from each other.
- It creates a smoother workplace by removing constraints, enabling transparency between departments, and engaging team members in most of the decisions that are made in the project.
- It enables a creative and innovative workplace where all members can share their ideas to innovate.

Working collaboratively is of great importance for your role as a solution architect. This is because when you collaborate with your team, you ensure the delivery of not only good architecture but also a good quality product at the end of the development.

Demonstrating influencing and negotiation skills

Strong influencing and negotiation skills are integral for the solution architect to resolve conflicts and develop win-win solutions. **Negotiating** is the ability to discuss a matter with a client or a team member and reach an agreement that is satisfactory for both of you. **Influencing** is a key part of a successful negotiation. It is the ability to negotiate and convince others so that they accept your suggestion.

Here are some essential negotiation skills that will support you in becoming an effective negotiator:

- Align the negotiation flow with your strategic goals for the short term and the long term.

- Before starting the negotiation, try to thoroughly prepare by collecting information about the people you are meeting with and the topics that you will negotiate.

- Prepare your **best alternative to a negotiated agreement** (**BATNA**). This is in case you do not reach an agreement in the negotiation at hand.

- Have the ability to be rational by separating personal issues and negotiation issues. This is important to see the opportunities and reach the objectives of the negotiation.

- Know how to form coalitions by discussing your ideas with potential allies who share common interests with you and who can influence other people involved in the negotiation.

- Build trust and reputation with the people you are negotiating with. You can do this by being respectful, transparent, and committed to your promises.

You will always be in a situation where you need to influence the decision-making peers involved in the solution you are designing. Keep developing your negotiation skills to build solid relationships, resolve conflicts, and make great deals.

Possessing a wide range of technical expertise

To be a true pillar of innovation, a solution architect should possess extensive technical expertise with the ability to consult management and engineering teams with technical recommendations. In particular, they need to be aware of the following:

- Software engineering and architectural design

- Information technology architecture, infrastructure, and cloud development

- Project management and product management

- DevOps tools and practices
- Business analysis

Breaking down problems efficiently

A big problem is hard to deal with; sometimes, it will make you feel overwhelmed. Approach the problem with a positive attitude and make it easier on yourself and the team by breaking it down into smaller issues that are easier to solve than the original big one.

Good architects can also minimize the occurrence of problems and tackle them before they occur. This can be achieved by doing good risk analyses and by keeping an eye on the details.

Being pragmatic

Every organization has its own set of standards, internal politics, deadlines, budget, and more. Good architects should be aware of the restrictions they might encounter during the implementation of a project. Therefore, they should deal with problems in a practical way, rather than by using an abstract theory that is not applicable in all cases. They should come up with realistic, timely, pragmatic, and efficient solutions that fit those restrictions. This is where decisions on business value-based architecture should be reached in order to optimize the company's **return on investment** (**ROI**).

So far, we have discussed some of the fundamental personality traits and key skills of the solution architect. In the next section, we will highlight a few common mistakes that you should deal with to avoid client dissatisfaction and any critical issues that could affect your project deliverables.

Taking a look at the common pitfalls for architects

Too often, mistakes are learned about the hard way. In this section, we are going to highlight some traps that can affect the outcome of the solution architect:

- Avoid over-architecting the solution; for instance, by adding unnecessary layers and components to the solution, which can instead be replaced with simple classes. Over-architecting will increase the development activities and complicate the troubleshooting in the case of errors, which will decrease the achievements that your clients are looking for. Try to simplify your architecture and make it more aligned with the business requirements.

- Consider writing custom code instead of using ready-made and open-source frameworks or libraries. For instance, let's assume you need to build a responsive web solution; here, you have two options.

  The first option is to reinvent the wheel and roll your own responsive framework. This option is costly since you need to develop the entire **CSS** and **JavaScript** from scratch. The second option is to use a popular open-source framework, such as **Bootstrap**, to build the responsive UI, which is a less costly option.

  Which option should you choose? Of course, in such a situation the best option is to use a reliable open-source framework to shorten the development time, to avoid errors that you will face if you develop your custom framework, and to make use of the powerful capabilities that are provided by the open-source framework. You need to learn when you should develop your own code and when to use an open-source framework by deciding which one is the best option for your solution.

- Jumping into development before planning or setting up the design will put the project deadlines, budgets, and goals at risk. Having a proper plan and design before starting the development is a must. It structures the work of your team members and sets expectations that should be aligned with the requirements. No matter how small or large the project you are working on, you should always follow the SDLC phases and adopt the output of each phase in order to move to the next one until you deliver the product.

- Not structuring the code in a way to make it reusable and extendable will result in you having a lot of repetitive code in many locations across the software, which performs the same task. This will increase the size of the source code and will complicate the maintenance tasks. This is because the code will become difficult to read and you might need to fix the same defect in many places. Try to keep your code **DRY (Don't Repeat Yourself)**, create components, make your software modular, and apply **Object-Oriented Programming (OOP)** when required by creating abstract classes and interfaces.

- Developing the product without paying attention to security will deliver a software solution that is open to security vulnerabilities. Security attacks can damage the system and its associated database, which can cause downtime and your client could lose money. Securing your software solution is no longer an afterthought but a foremost one. Paying attention to security during the design and development of the software helps you avoid malware and prevent hacking attempts.

- Avoid relying on traditional development methodologies to build software rather than applying Agile software development methodology. Agile methodology is widely used by software development companies so that they can manage their projects effectively. It is a modern practice that adds value to your team and to the solution you are developing.

  Agile is an evolutionary process that promotes a high level of collaboration between team members from different departments by bringing them all together to deliver the project. It increases the productivity of the team, allowing for multiple deliverables in a short time. Additionally, it allows clients to closely contribute to each stage by providing their feedback, which prevents any disappointments at the end of the project.

- Avoid not paying enough attention to code optimization and performance.

- Avoid not spending enough time on **User Acceptance Testing** (**UAT**). All types of testing are very important in order to deliver high-quality products. UAT is a type of testing performed by the client to decide whether the requirements have been met. This is before moving to a production/live environment.

  Without a proper UAT phase, the product could be rejected by end users for many reasons, such as bad performance, missing features, and the product being buggy. To avoid this situation, make sure you prepare, in advance, the test cases based on the use cases that were agreed with the client. These test cases will be used in the UAT phase to verify every single functionality.

- Avoid not allocating team members with the right skillsets that should be aligned with the project needs. Remember that success isn't possible without the right team members.

- Avoid the a of good planning by having an unrealistic timeline and budget and finding out about this at a late stage of the project. This can create conflicts, especially if you are working on multiple projects that are running in parallel. You must have a realistic plan with clear deadlines and goals; this should be identified during the planning phase of the project. You can apply the Agile methodology to avoid this pitfall.

These various pitfalls can occur regularly and probably in conjunction with each other, which could affect the quality of your deliverables and the reputation you have earned with your clients. We have described a few of the many pitfalls that you might experience. Being aware of these pitfalls will prepare you so that you can overcome them. As a solution architect, you need to constantly solve problems, try new things, and help align a project with the company vision and values.

In the next section, we will explore the difference between the enterprise architect, the technical architect, and the solution architect.

The enterprise architect versus the technical architect versus the solution architect

There are three different architecture-related roles in the information technology industry. Each of these roles is equally essential in the software development life cycle and cannot be replaced by any other positions:

- **Enterprise architect**
- **Technical architect**
- **Solution architect**

Now, let's learn more about the difference between these roles:

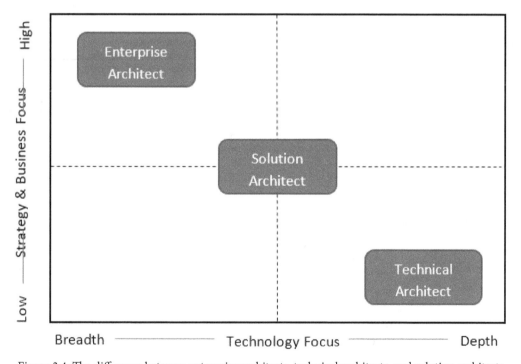

Figure 3.4: The difference between enterprise architects, technical architects, and solution architects

Enterprise architects are responsible for collaborating with key stakeholders to define business goals and establishing the entire enterprise infrastructure (such as software and hardware), which supports the needs of the organization. They mainly focus on implementing and managing complex IT solutions that target critical and strategic business goals at the same time.

Meanwhile, **solution architects** have a practical role to play within the organization. They collect the business requirements, then analyze them, and finally, turn them into a new software solution that uses the company's standards and technology stack.

Technical architects mainly oversee the technical architecture of the solution and the core technology used in the implementation. Their main responsibility is to provide technical leadership to the development team and decide on every technical aspect of the software solution.

The following diagram depicts the structure of the enterprise architecture stack in an enterprise that provides digital transformation services:

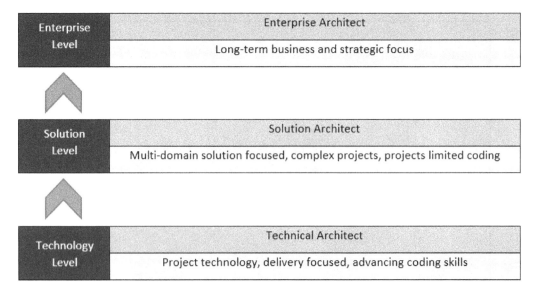

Figure 3.5: The enterprise architecture stack

By having a solution architect onboard, companies are able to primarily create a solid structure that aligns their corporate vision and goals with various technologies.

Summary

In this chapter, we provided a quick introduction to what solution architecture is. Then, we learned about some essential personality traits and soft skills that are required to become an effective solution architect. Later in this chapter, we highlighted few common pitfalls that should be avoided during the project development life cycle. Finally, you learned about the differences between enterprise architects, technical architects, and solution architects.

In the next chapter, we will dig deep into the principles of solution architecture, and you will learn about the seven popular UML diagrams that will help you to model your solution architecture.

Section 2: Designing a Solution Architecture

In this section, we will discuss the key principles of solution architecture and we will delve into the most frequently used UML diagrams with concrete examples. Then, we will get to know the process of creating and designing an architecture with UML.

After that, we will explore the key architecture patterns and how to choose the right pattern for our solution. Moreover, we will explore the design and runtime quality attributes of the solution architecture. Later in this section, we will dig deep into security considerations and how to secure our ASP.NET solutions, and then we will discuss the main types of testing and what the best practice is in this regard.

This section comprises the following chapters:

- *Chapter 4, Designing a Solution Architecture*
- *Chapter 5, Exploring Architecture Design Patterns*
- *Chapter 6, Architecture Considerations*
- *Chapter 7, Securing ASP.NET Web Applications*
- *Chapter 8, Testing in Solution Architecture*

4

Designing a Solution Architecture

In the previous chapter, we learned about some of the essential traits and skills that are needed to build your potential and become an effective solution architect. We also looked at a quick overview of what solution architecture is.

In this chapter, we'll begin to focus more on solution architecture practices. In particular, we'll take a look at the key principles of solution architecture, and we'll explore popular **Unified Modeling Language** (**UML**) diagrams that are recommended for designing medium- to large-scale solutions.

In this chapter, we will cover the following topics:

- Exploring the key principles of solution architecture
- Delving into the most frequently used UML diagrams with concrete examples
- Walking through the process of creating a design architecture with UML

By the end of this chapter, you will have enriched your knowledge and understanding of popular UML diagrams, and you will have learned how to use them in your design. Additionally, you will learn about the key principles of solution architecture and how they can influence your design process.

Exploring the key principles of solution architecture

Architecture principles outline the fundamental procedures and guidelines that are required to design, build, and deploy a successful software solution. They are meant to influence your architecture approach and improve the quality attributes of the solution.

There are many principles out there that we can adopt in our methodology of work to prepare the design architecture. We can even define our own principles if we think they will add value to the architecture design or if we think they will efficiently improve the design process. Most importantly, we need to make sure we offer a good balance between theory and practice and that we adopt practical and powerful principles that will drive the business and technical changes in our solution's architectures.

In general, we should aim for between 10 and 20 guiding principles for our solution architecture practices. Make sure that you do not have too many principles. This is because they become hard to remember and difficult to apply, which will limit our architecture's flexibility. In such cases, it is better to keep them simple and focused.

There is a standard way and recommended format that you can use to define an architecture principle. Usually, a principle is divided into four main parts:

- **Name**: The name should reflect the core value of the principle. It should be simple and easy to remember.

- **Description**: The description is a statement that clearly defines and explains the principle.

- **Rationale**: The rationale is a statement that highlights the business benefits of obeying the principle. It can also explain the correlation with other principles.

- **Implications**: The implications should highlight the technical requirements and business requirements that are needed to adopt this principle.

These elements are meant to support the understanding of each principle and justify its usage in the solution architecture.

Any principle we adopt should fall into a specific category or domain. According to **The Open Group Architecture Framework** (**TOGAF**), architecture principles broadly fall into four domains:

- **Business principles**: These are a set of guidelines that focus on the business aspects of the solution.

- **Data principles**: These define the standard guidelines to manage and structure data. Additionally, they enforce a set of security measures in which to protect the solution assets.

- **Application principles**: These deal with the attributes of an application such as performance, user experience, and how modules or subsystems interact with each other.

- **Technology principles**: These elaborate on the technical guidelines and requirements that are necessary for the success and continuity of the solution.

> **Tip:**
>
> TOGAF is a framework and methodology that has been developed to provide a high-level approach to design and also build enterprise information technology architecture. You can learn more about architecture principles, as defined by TOGAF, at `https://pubs.opengroup.org/architecture/togaf9-doc/arch/chap20.html`

Here is an illustration that highlights the key principles of solution architecture. You can see that the four domains are consolidating a set of design guidelines in different domains to produce a solution that is flexible, scalable, and reusable:

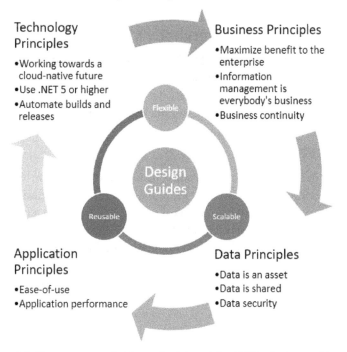

Figure 4.1: The recommended key principles of the .NET solution architecture

In the upcoming sections, we will explore the preceding principles that you really need to know. We can adopt these principles as is, or we can add, modify, and remove some principles based on our needs. However, always remember that a principle is made to benefit the solution we are suggesting, not to add any obstacles or complications.

Business principles

The solution architecture, including all the development deliverables and their quality, is of critical importance. However, we shouldn't only focus on the project plan, schedule, and outcome. Remember that the reason you are building the solution is to solve business problems. Therefore, it is a must for the solution architecture to align with the business goals and objectives. In the following sections, we will discuss three business principles that should be taken into account.

Maximizing the benefit to the enterprise

All architectural and information management decisions must be made in a way that ensures the maximum number of benefits to the entire enterprise. For instance, the solution should bring long-term values to all entities within the enterprise, not solely to one department or minor group. This principle encourages the high-performing collaboration of service above self.

Information management is everybody's business

All key stakeholders, business experts, and technical teams are responsible for coordinating together as one committee to define the business objectives of the solution and ensuring that they align with the enterprise goals. Essentially, everyone in this committee is responsible for doing their own part in building the solution and managing it.

Business continuity

In the case of system failure, the solution architecture should ensure the business continuity of the enterprise. Put simply, any kind of system failure, including hardware, software, and data corruption should not affect the continuity of the business activities and operations. For instance, the solution design should suggest a state-of-the-art recovery mechanism, system redundancy, or failover backup plan to smooth the operation of the business functions in the case of disasters. The key stakeholders should define the criticality of the solution to the enterprise operations and decide what type of failover plan should be applied to ensure business continuity.

Data principles

Data is an essential part of business processes; it is a valuable asset that empowers enterprise stakeholders to make strategic decisions based on key metrics and performance indicators. There are three data principles that should be taken into account when designing and building a software solution. We will explore each of them next.

Data is an asset

Although many architects know about this principle already, we still find that data is not considered with a high level of importance in the way it should be. Data is a core business asset of the organization. Therefore, the design of a solution should ensure the proper storing, managing, and retrieving of the data with high-security measures for better protection.

Data is shared

Accurate data that is stored in a centralized repository is the backbone of the software solution, and timely access to the data is very important to improve the efficiency of the decision-makers who are using the solution. Additionally, business users need data in order to perform their daily duties. Therefore, your solution design should allow timely access to the data based on the access rights of your users.

Data security

The software solution should ensure the integrity and confidentiality of the data. It should protect the data and prohibit unauthorized access or unlawful processing. There are many data policies out there to protect the data; we need to comply with these policies based on our target users. For instance, if we are building a software solution targeting European citizens, then complying with the **General Data Protection Regulation (GDPR)** is a must.

Application principles

The guiding principles of modern applications are evolving. They should be dynamic principles; remember that what was applicable in the past is not relevant in today's architecture. Always look to improve the principles that you adopt and make it a continuous phenomenon. Here are the principles that really matter and that make a difference in the delivery of modern and efficient applications.

Ease of use

The application should be user-friendly, easy to use, and visually appealing. We should embrace simplicity; that is because the easier an application is to use the higher the chance that it will be adopted by our end users. Always put the users first; they are looking to use applications that can facilitate their work and make it efficient in a short period without having to spend a long time learning about it before they start getting value from it.

Optimized application speed

We live in the era of digital transformation; users are looking for real-time response applications. Therefore, the speed of the application is an important factor to consider, as it can affect the entire user experience.

Technology principles

Our technology principles should always follow the latest technology trends. It is very important to modernize your company's technology platforms and development practices. This will allow you to design modern digital solutions. Let's explore the three principles that should be adopted in any .NET development team. We have listed them next.

Working toward a cloud-native future

Cloud-native applications have proven to be the future of software. The solution we are trying to build can benefit from the platforms, services, and processes that are hosted in the cloud. For instance, the Azure services are highly scalable, easy to modify, and connected, which allows us to extend the application's capabilities with less coding.

Using .NET Core (.NET 5 or later)

The latest release of .NET Core is called **.NET 5**. It is a free and open source framework that can be used to build any type of application. It is a cross-platform framework that has inherited all the significant advantages of the regular .NET Framework. One of the key improvements of .NET Core is the performance, so consider using this framework when you are building new solutions.

Automating repetitive development tasks

Preparing the release of your solution can be time-consuming, particularly if you are aiming to release several builds and hotfixes. You can plan out automated jobs to minimize the manual intervention needed within such tasks, make use of DevOps' tools to automate your builds, and test plans, too.

In this section, we learned about the key principles that should be adopted in your solution architecture. These principles provide guidelines for four primary aspects of the solution: business, data, application, and technology. Applying these principles will give you the ability to deliver a solid solution that is scalable, reusable, and easy to maintain.

In the next section, we will explore six popular UML diagrams that you should use when designing a software solution.

Learning to model software architecture using UML

UML is a standard graphical representation that allows us to visualize the specification and design architecture of the software solution; it is a simplified way in which to communicate our architecture to the solution stakeholders. The purpose of the UML is to provide the development team and the business analysts with a unified design modeling notation that empowers them to explain complex business processes with simplified diagrams. Additionally, it enables us to construct and visualize the different software components and how they relate together, which defines the entire design architecture of the solution.

There are two categories of UML diagrams: **structural** and **behavioral**. Structural diagrams emphasize the static view of the system. They are used to visualize the different components and objects of the software. Mainly, they describe what is contained in a system.

Behavioral diagrams emphasize the dynamic view of the system. They are used to visualize the business specifications by describing the processes and functionalities supported by the software. Primarily, they describe what must happen in a system.

The following diagram shows the different types of UML diagrams:

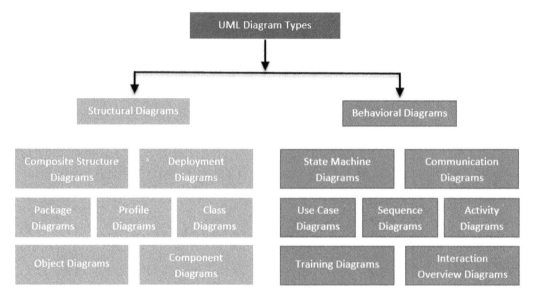

Figure 4.2: A list of UML diagram types

In the upcoming sections, we will explore the most frequently used diagrams. Although there are many different types of UML diagrams that we can use to model solution architecture or describe system functionalities, we will explore the popular UML diagrams that are frequently used by most architects to document the different aspects of a software solution. They are listed as follows:

- Component diagrams
- Class diagrams
- Sequence diagrams
- State diagrams
- Activity diagrams
- Package diagrams
- Use case diagrams

We will learn when to use each diagram, and we will also explore the different notations and symbols of each. Then, we will examine an example of each diagram. Let's start learning about these diagrams next.

Component diagrams

The UML component diagram is used to graphically represent the different modules and components in the software system, including the relationship and interaction between these modules. A module is a set of classes or interfaces that provides different functionalities but are grouped into one business routine.

The benefits of component diagrams

Component diagrams can help you by doing the following:

- Visualizing the overall physical structure of the software system
- Describing the system's components and how they are related
- Grouping the object-oriented classes based on a common service objective
- Modeling the .NET source code or the database of the system

The notations and symbols of a component diagram

Here are the different shapes and symbols used to draw a component diagram:

Element	Description	Symbol
System Component	Used to represent a logical block of the system.	 Component
Interface	Used to represent a group of operations provided or required by the component.	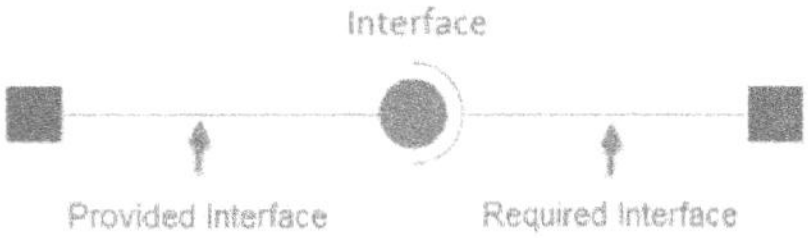 Interface — Provided Interface — Required Interface
Dependency	Used to visualize the dependencies between the different components of the system.	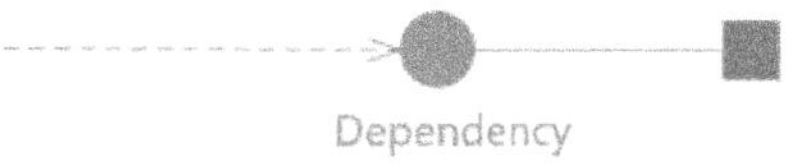 Dependency

Figure 4.3: The notations and symbols of a component diagram

The component diagram of a shopping system

Let's assume that we want to build a simple online shopping solution. We will use a component diagram to describe the different components in this system.

69

First, we need to identify these components by grouping the functional requirements according to their purpose. In this example, we have three components:

- The orders
- The customer accounts
- The products inventory

Next, we need to identify and visualize the relationships between these components. To make an order, the customer should provide the necessary input to finalize the order. That's why we need to use the interface symbol to relate the order to the customer; the customer should then select one or more products from the inventory. This will ensure that an order is fully associated with the products. So, we will use the dependency symbol to relate the order to the products. Here is an example of a component diagram describing the three components along with their interactions:

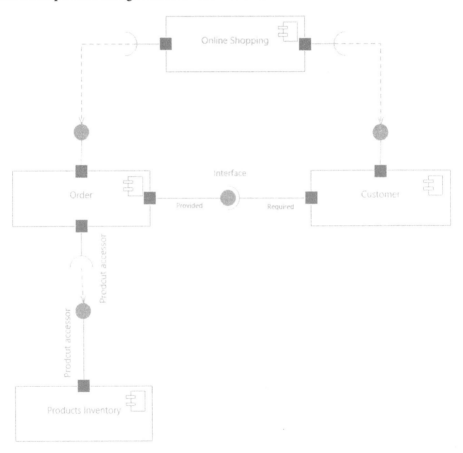

Figure 4.4: A UML component diagram for online shopping

Note that you can use the component diagram to illustrate the physical files in your source code. This is very helpful when doing forward or reverse engineering to identify the set of source code files. Additionally, you can use the component diagram to model a physical database.

Class diagrams

The UML class diagram is used to describe the structure of the object-oriented system by graphically representing the classes with their attributes and operations, including the relationships between these classes.

The benefits of class diagrams

Class diagrams are very popular among software engineers. They are very powerful and can be beneficial when you want to describe the object-oriented classes within the software system. We can use UML class diagrams to do the following:

- Describe each class in the system with its structural features (attributes) along with its behavioral futures (operations).

- Draw the relationships between classes, such as **abstraction** and **association**.

- Visualize the data models of the system.

- Create a detailed model of the software from a business perspective, which is very helpful for non-technical stakeholders in order to understand the general overview of the system.

- Generate **C#** source code from the class diagrams.

The notations and symbols of a class diagram

Class diagrams are simple and easy to read. Here is an example class diagram to help you understand the different notations:

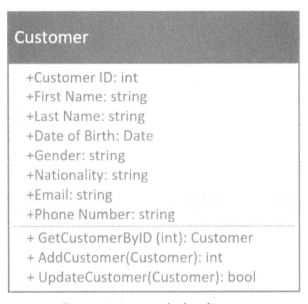

Figure 4.5: An example class diagram

As you can see, the standard class notation is composed of three sections:

- The upper section contains the name of the class.
- The middle section contains the class attributes/members with their types.
- The bottom section contains the class methods displayed in a list format.

There are different visibility symbols that are used to indicate the access level of information contained in a class:

Symbol	Access modifier	Description
+	public	Objects marked with (+) are accessible from everywhere in the class library or assembly.
-	private	Objects marked with (-) are only accessible from within the class itself, they are hidden from outside the class.
#	protected	Objects are accessible from within the class and all inherited/derived classes.
~	internal	This access modifier is specific to the C# language, whereas UML is a universal modeling language. Therefore, you can decide to use (~) to indicate that these objects are accessible only from within classes in the same assembly.
underlined	static	Objects that are underlined have only one copy, the underline indicates that this is a static object.

Figure 4.6: C# access modifiers and their symbols in UML

We can use cardinality notations to define the type of relationship between two classes. For example, one customer can have one or more orders (that is, one to many relationships), while another order can have one customer (that is, one to one). The following table shows the different symbols of cardinality:

1...1	One to one
1...n	One to many
...	Many to many

Figure 4.7: Cardinality types

There are different types of relationships between classes. The following table shows the symbols of these relationships:

Element	Description	Symbol
Association	This is the default relationship between two classes. It indicates that one class is aware of another class, they see each other.	
Inheritance	It is called generalization as well. The symbol indicates that a class inherits the features of another class. This is a kind of "is a" relationship. For example, a Mercedes is a type of car.	
Realization	It connects a client element with a supplier element, it indicates that the client should implement the behavior defined in the supplier. For example, the student realizes the specifications provided by the person class.	
Dependency	It indicates that one class depends on (uses) the other class. For example, a class "Project Manager" depends on class "Project".	
Aggregation	It is a kind of "has a" relationship. For example, a class "Team" has a "Member".	
Composition	It is a kind of "is-composed-of" relationship. For example, a class "Car" is composed of a class "Engine".	

Figure 4.8: The relationship types between classes

The class diagram of an online shopping system

In *The component diagram of an online shopping system* section, we visualized the component diagram for an online shopping system. In this section, we will illustrate the class diagram of this system:

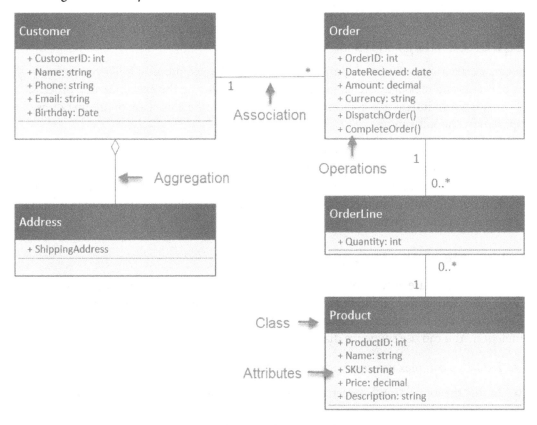

Figure 4.9: A class diagram of an online shopping system

As you can see, each box represents a class in the header. We should provide the class name and then list all the properties and operations of the class. The interaction between the classes is visualized through the lines. We call them relationships.

The class diagram can be used during the entire life cycle of the software, starting with visualizing the high-level conceptual idea of the software, then when creating a detailed understanding of the specifications, and, finally, during the implementation of the software. The class diagram is an essential modeling technique that is used to visualize all the object-oriented objects in your system, so make sure you master it.

Sequence diagrams

A sequence diagram is used to document a business or logical process. It illustrates the flow of events and messages exchanged between objects during the execution of a process. It is considered an interactive diagram because it can describe a use case or an operation supported by the software.

This interactive operation can happen between a user and the software you are building, between your software and other systems (such as middleware integrations), or between the sub-modules within the same software. For example, we can use the sequence diagram to explain the authentication mechanism in our system or to illustrate the booking process of a hotel reservation system.

The benefits of sequence diagrams

Sequence diagrams are very helpful when it comes to describing a complex operation or a use case. You simply highlight the objects involved, the order of the steps in the operation, and the messages exchanged from the beginning of the process until completion. You can use sequence diagrams if you want to do the following:

- Explain a complex use case with several steps.

- Model the interaction between objects and components during an operation.

- Illustrate an integrated procedure between your system and another third-party system.

The notations and symbols of a sequence diagram

The following table lists the basic notations and symbols that you must know in order to create a sequence diagram:

Element	Description	Symbol
Object Lifeline	Represents an object or component. The vertical dashed line represents the sequence of actions that occurs during an interaction, while time progresses downward from top to bottom	
Actor Lifeline	Represents a type of role played by an entity that is external to the system. The vertical dashed line represents the sequence of actions that occurs during an interaction, while time progresses downward from top to bottom.	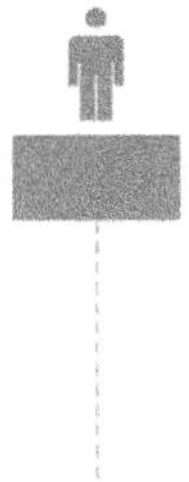
Activation	It is added to a lifeline. It represents the period when the participant is executing an operation.	
Call Message	The call message is a kind of message that is passed during an operation from the invoker to the target lifeline.	
Return Message	The return message is a kind of message that is returned during an operation from the target lifeline back to the invoker.	
Self-Message	The self-message is a kind of message that is passed during an operation from a lifeline to itself.	
Loop	Indicates a loop. Use this notation to indicate the condition under which you want to repeat the action.	 loop [parameters]

Figure 4.10: The notations and symbols of a sequence diagram

77

In the next section, we'll create a sequence diagram.

The sequence diagram of a shopping cart

In the following example, we have illustrated a simple sequence diagram of an example use case for an online shopping process. The diagram includes these lifelines:

- The customer who wants to shop from the online system.
- The shopping cart interface, which holds the items that a user wants to buy.
- The order module, which processes the user request and confirms the payment.

The process is described using these sequence messages:

- Users can add a product to the shopping cart.
- Users can remove a product from the shopping cart.
- Users can adjust the number of items.
- Users can see the total price of the selected items.
- Users can confirm the order.

The following diagram contains a loop fragment that allows the user to add more products or items to the shopping cart before confirming the final order:

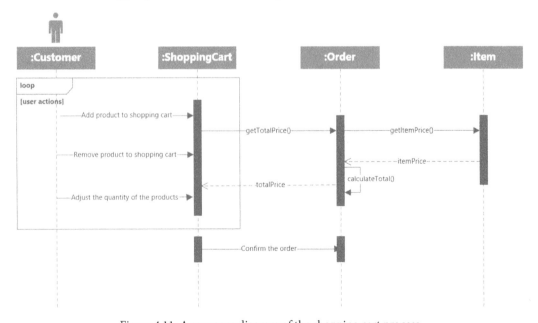

Figure 4.11: A sequence diagram of the shopping cart process

In the next section, we'll learn about state diagrams.

State diagrams

A state diagram (or a state machine diagram) is typically convenient when you want to describe how your system behaves and responds. It is a combination of states, transitions, events, and activities. It is used to model the process of a particular function and shows all of the transitions from one state to another. It can also describe a single object and illustrate how that object behaves in response to a series of events in your system. The state of an entity is defined by the values of its attributes, which are controlled by a particular event at a specific time.

State diagrams are very useful when you wish to model the behavior of an interface, class, or collaboration, and the business processes triggered by specific events. It also helps you improve processes by eliminating unnecessary steps and identifying missing steps that should be added to the process to make it more efficient.

The benefits of state diagrams

State diagrams are used to describe how an event can change the behavior of a process. We can use state diagrams to do the following:

- Visualize the dynamic view of a system.

- Model the flow of states in a business process scenario.

- Explain an event-driven process using the state of objects when they move from one step to another.

- Illustrate interactive functionalities in the system.

The notations and symbols of a state diagram

The following table lists the basic notations and symbols that you must know in order to create a state diagram:

Element	Description	Symbol
Initial State	It represents the first state in the process.	
State box	It represents a particular state in the lifespan of the process.	 State Title
Decision box	The decision evaluates the previous transition to decide which path to take to complete the process.	
Final State	It represents the final state in the process.	
Transition	It indicates the progress of the process from one state to another.	

Figure 4.12: The notations and symbols of a state diagram

In the next section, we will use these symbols to draw a state diagram to describe a two-factor authentication process.

The state diagram of a two-factor authentication process

In the following diagram, we have described the events of a two-factor authentication process along with the transitions from one state to another:

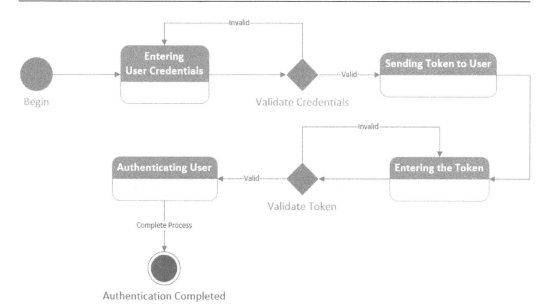

Figure 4.13: A state diagram describing an authentication process

The preceding diagram shows the first event that occurs as a result of the user providing their login credentials. Then, the system validates the credentials and sends the authentication token back to the user if the credentials are valid; otherwise, the user is requested to enter the valid credentials again. After that, the user should submit the token for verification by the system, which will decide to complete the login process if the token is valid; otherwise, the user is requested to re-enter the token to continue.

Activity diagrams

One of the important diagrams in UML is the activity diagram. It is similar to the state diagram in terms of illustrating the dynamic view of the system. An activity diagram is a flowchart that describes the flow of control from one activity to another activity among the objects in a system. It is mostly used to model business processes. The activities described in an activity diagram can be sequential and concurrent.

The benefits of activity diagrams

An activity diagram is a useful flowchart that describes the activities performed by a process in our system. Activity diagrams can help us to do the following:

- Explain the steps of a use case scenario by describing all of the activities.
- Learn the logic of a particular algorithm.
- Brainstorm and model business processes and workflows.

The notations and symbols of an activity diagram

Before you begin creating an activity diagram, you must understand its notations and symbols. The following table lists the main symbols of an activity diagram:

Element	Description	Symbol
Initial Node	Represents the starting point of the process.	
Activity	Represents an action in the process.	
Connector	Represents the control flow from one activity to another.	
Join Node	Represents the end of two parallel activities by combining them into one single activity.	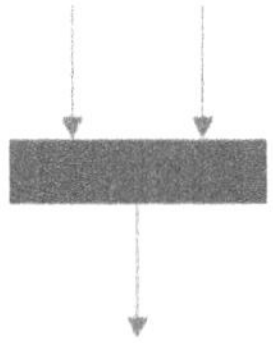
Fork Node	Represents the beginning of a parallel activity by splitting one activity into two concurrent activities.	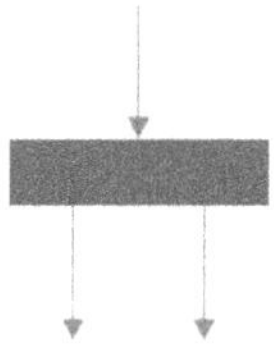
Decision	Represents a decision in the flow which will lead to alternate paths.	
Note	It is used to clarify certain activities or flow by adding additional comments to the diagram.	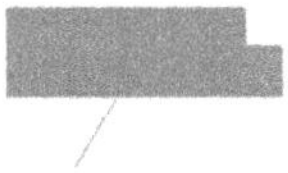
Swimlane	It is used to group the activities logically.	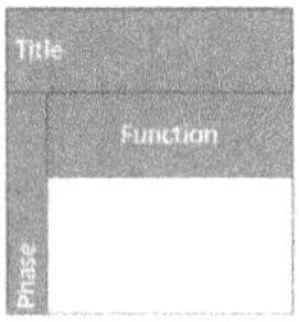 Title / Function / Phase
Final Node	It represents the completion of the process.	

Figure 4.14: The notations and symbols of an activity diagram

In the next section, we'll take a look at an example of an activity diagram.

The activity diagram of an ATM system

In the following example, we have described a basic process for an ATM system using an activity diagram:

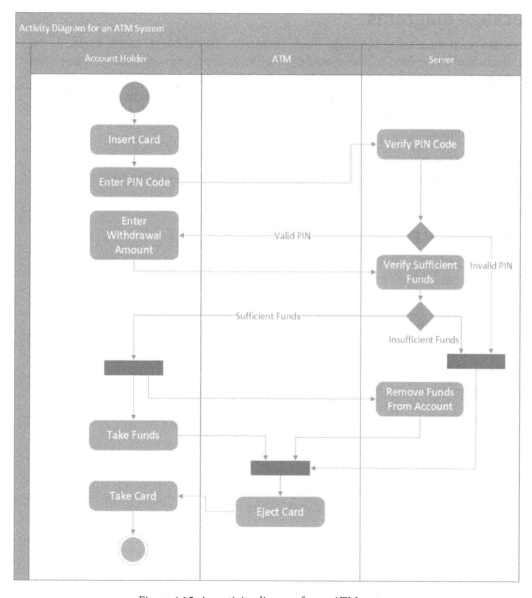

Figure 4.15: An activity diagram for an ATM system

As you can see, the preceding diagram describes the process of an ATM transaction. It starts by inserting the card and then providing the PIN code. Initially, the server will verify the PIN code. If it is valid, then the server will advise the ATM to allow the customer to proceed, enter the withdrawal amount, and complete the transaction; otherwise, the card will be ejected.

Package diagrams

The main purpose of using a package diagram is to describe the high-level logical architecture of our system by visualizing the various layers along with their dependencies in which a layer represents a group of classes. The physical components in the system are grouped into layers based on their roles and the tasks they perform in the system. It is possible to create nested layers within a single layer to describe the advanced details of a major component.

After defining the main layers in your system, we need to visualize the dependencies or relationships between the layers. This will describe the interactions that are occurring between layers.

The benefits of package diagrams

By unifying the major components into layers, a package diagram can make our architecture easy to understand. We can use the package diagram to help us perform the following tasks:

- Explain the high-level logical architecture and structure of the system.
- Visualize the major components or functional units of the design and their interdependencies.
- Identify the possibilities of integrating your system with a third-party system.
- Discover gaps in the architecture that could prevent your system from evolving.

- Communicate the effort that is required in the case of a major change that might affect several layers.

- Align with the development team on the intended architecture. This diagram provides you with the ability to compare your architecture with what is being implemented during the development.

The notations and symbols of a package diagram

The following table lists the main symbols of the package diagram:

Element	Description	Symbol
Layer	A logical grouping of the physical components such as classes, projects, web services, etc.	
Dependency	Represents the relationship between two layers.	

Figure 4.16: The notations and symbols of a layer diagram

In the following section, we'll examine the package diagram of an ASP.NET web solution.

The package diagram of an ASP.NET web solution

In the following diagram, we have used a package diagram to describe the high-level architecture of a typical ASP.NET solution without mentioning any details about the classes and assets within each package:

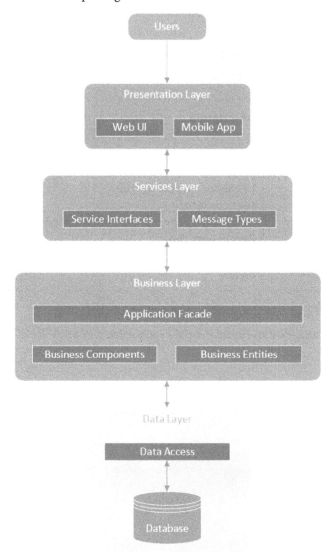

Figure 4.17: A package diagram for a typical ASP.NET solution

In the preceding diagram, we have represented the packages within the multitier architecture of an ASP.NET solution. The **Presentation Layer** includes the user interface of the solution, and the **Services Layer** represents a middleware service that provides high-level communication services to web and mobile apps. The **Business Layer** holds all the business logic and entities. As for the **Data Layer**, this includes methods in which to access data stored in the database.

Use case diagrams

One of the most exciting moments in software engineering is the point at which the product you are designing and developing meets the customer's needs. There is nothing better than having clear use cases to reach this target. Use cases are essential for describing the product from the client's perspective. There are two types of use cases: **textual** and **visual**.

In *Chapter 1, Principles of the Software Development Life Cycle*, we learned about textual use cases and how to prepare them. In this section, we are going to cover the visual representations of use cases.

A use case diagram is used to visualize the user's requirements; more specifically, it is used to visualize the system behavior and the interaction between the users and the system.

The benefits of use case diagrams

A use case diagram is a simple and effective technique that can be used to visualize the user's interaction with the system. It doesn't show all the detailed user requirements but only the interaction of the use case. Use case diagrams can help you by doing the following:

- Illustrating the users' interaction with the software
- Visualizing the functional needs and the scope of the system
- Showing the high-level steps of a use case
- Aligning the user's requirements with the implementation and supporting the generation of the test cases

The notations and symbols of a use case diagram

The notations of a use case diagram are simple and straightforward. The following table lists the main symbols of a use case diagram:

Element	Description	Symbol
Actor	It represents a user with a specific business role who is interacting with the system.	 Actor
Use Case	It represents a coherent functionality unit provided by the system.	 Use Case
Subsystem	It represents a system or subsystem that can contain multiple use cases.	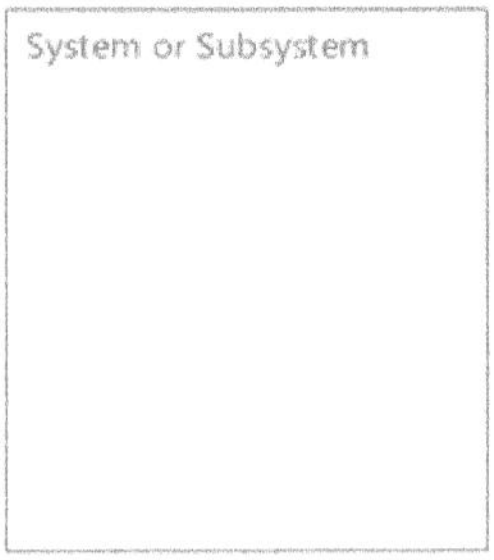 System or Subsystem
Association	It represents a communication link to connect an actor to a use case.	

Figure 4.18: The notations and symbols of a use case diagram

A use case diagram for the interaction between the customer and the ATM

In the following diagram, we created a use case diagram to describe the main functionalities supported by the ATM system, including the interaction with the customer:

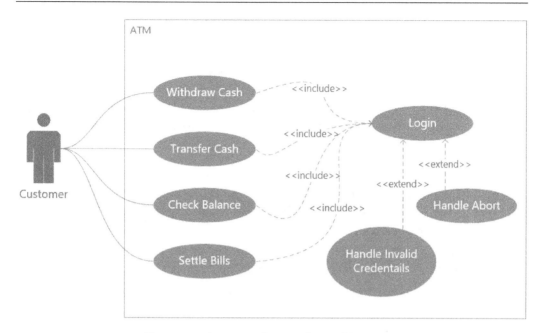

Figure 4.19: A use case diagram for an ATM system

In the preceding diagram, the actor is the customer who is using the ATM system and trying to authenticate to complete a transaction. The colored ellipses represent the functionalities supported by the system, and the lines represent the association and interaction between the blocks.

In this section, we learned about the most frequently used UML diagrams that are essential to document many aspects of the solution such as the solution design, the structure of the object-oriented system, and the business and logical processes. We also learned about how our system behaves and responds to user interactions, the flow of control from one activity to another activity among the objects in the system, the high-level logical layers of your system with their dependencies, and the requirements from the client's perspective.

In the next section, we will learn how to create a solution architecture using UML.

Designing architecture with UML

So far, we have learned that a UML diagram is a single simplified representation of the software. We will need to build various UML diagrams in order to understand the complete aspects of the system and to communicate our architecture design to stakeholders and different types of users. Grouping these UML diagrams into logical subsets will create a particular view of the system. The architecture design is represented in a collection of five views. These views are as follows:

- **Use case view**: The **use case view** represents the focal point for all of the other views because it includes the user requirements, including all of the system functionalities. Without this view, you cannot build the other views.

- **Design view**: The **design view** is intended to illustrate how the functionality defined in the use case view is designed inside the system in terms of classes and their relationships. This view is mainly described by the UML class diagram.

- **Implementation view**: The **implementation view** describes the core components of the system and the interaction between them. It is mainly represented by the UML component diagram.

- **Process view**: The **process view** describes the flow supported by the system. It does this by converging the performance, scalability, and throughput of the system. It is mainly represented by the UML class diagram and is similar to the **design view**, but focuses on the active classes involved in the processes.

- **Deployment view**: The **deployment view** describes the deployment of the system by focusing on the system topology, distribution, delivery, and installation. It is represented by the UML deployment diagram.

The following diagram illustrates the five views along with the use case view in the center, which is connected to all of the other views:

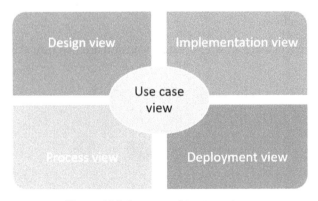

Figure 4.20: System architecture views

The following table summarizes the architecture views, along with the related UML diagrams, to help in the understanding of the different aspects of the system:

View	Stakeholders	Static Aspects	Dynamic Aspects
Use case view	End users Analysts Testers	Use case diagrams	Interaction diagrams State diagrams Activity diagrams
Design view	End users Developers	Class diagrams Package diagrams	Interaction diagrams State diagrams Activity diagrams
Implementation view	Developers Configuration managers	Component diagrams	Interaction diagrams State diagrams Activity diagrams
Process view	Integrators	Class diagrams	Interaction diagrams State diagrams Activity diagrams
Deployment view	System engineer	Deployment diagrams	Interaction diagrams State diagrams Activity Diagrams

Figure 4.21: System architecture views along with their related UML diagrams

The solution architect is responsible for creating the initial version of these views with their related diagrams and handling the updates that construct the solution architecture. The entire solution architecture is used to influence and guide the development activities throughout the project life cycle.

Summary

In this chapter, we explored the key principles that outline the fundamental procedures and guidelines required to design, build, and deploy a software solution. Additionally, we learned about the essential UML diagrams with real examples to illustrate the different elements in each diagram, along with their benefits, and when to use each of them. Later in this chapter, we explored the process that is involved when constructing a solution architecture with UML.

In the next chapter, we will dig deep into the core architecture patterns. We will focus on the microservices architecture, and we will learn how to choose the right pattern for a specific solution.

5
Exploring Architecture Design Patterns

In the previous chapter, we learned about the key principles of solution architecture. We also explored the most frequently used UML diagrams and when we should use each one to create a view of the solution architecture.

In this chapter, you will learn about the top architecture patterns that you must know to build a solid software architecture.

In this chapter, we will cover the following topics:

- Introducing the architectural patterns
- Exploring key architecture patterns
- Learning how to choose the right pattern for your product

By the end of this chapter, you will understand architectural patterns. We will enrich our knowledge by exploring the top architecture patterns with example use cases. Additionally, we will explain the criteria that we should focus on when choosing the right architecture pattern for our software product.

Before we begin learning about these patterns, first, let's understand what an architectural pattern actually is.

Introducing the architectural patterns

An architectural pattern is a reusable solution architecture to a common problem that we might face in different business industries and on various occasions. It offers predefined guidelines along with a set of rules to establish the underlying structure of the solution.

It is important not to mix up the **.NET** design patterns and the architectural patterns. The first one represents a way in which to organize classes to make your source code more reliable, scalable, and easy to maintain, which will solve various problems that are internal to a specific component or module in our system. In comparison, the second one has a broader scope within the entire solution as it defines the high-level abstract structure of the solution. As a solution architect, you must possess knowledge of both types of patterns:

- **Design Patterns** develop classes with object-oriented principles.
- **Architectural Patterns** help to define and maintain the overall structure of the entire system.

The following diagram shows the different levels of architectural decisions that you might need to make as a solution architect:

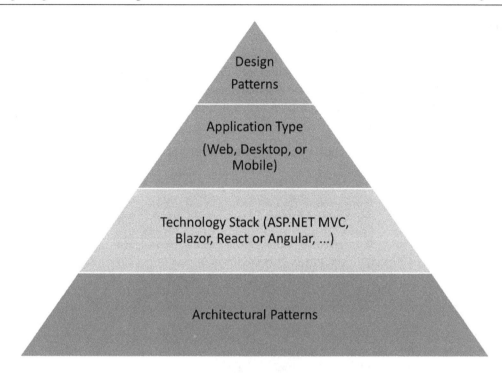

Figure 5.1: The various levels of architectural decisions

In the preceding diagram, you can see that choosing the architectural pattern is one of the earliest decisions that you must take. Following this, you should choose the right technology stack that you will use to build the product. After that, you need to decide upon the type of application and the design patterns that will help you to organize the code and make it both reusable and extensible.

Now, let's begin by getting to know the most popular architecture patterns.

Popular architecture patterns

In this section, we will explore five popular architectural patterns. We will explain the core concept of each pattern, and then we will outline the key components of each architecture pattern. This should help you learn about the usefulness of architecture patterns and support you in choosing the right pattern for a proposed solution. Let's begin with a layered architecture.

Layered architecture

This type of architecture is widely known by most architects and developers as **n-tier architecture**. It is used to structure the system into different layers, where each layer consists of a set of classes grouped in one assembly based on a specific context. The layers are structured horizontally so that each layer can consume services from one layer or the many layers that are beneath it.

In most cases, this architecture consists of three main layers, as shown in the following diagram:

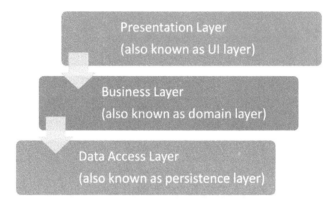

Figure 5.2: A typical 3-tier layered architecture

As you can see from the preceding diagram, these are the three main layers:

- **Presentation Layer**: This layer represents the component that is responsible for handling all user interactions through pages, menus, buttons, links, reports, forms, and more. It contains all the graphical designs and defines what the application looks like. It is the only layer that is visible to end users.

- **Business Layer**: This contains the business logic, business rules, and entities that define the behavior of the entire solution.

- **Data Access Layer**: This contains the code responsible for manipulating the database layer, which is where all the data is stored (for example, **SQL Server**, **Oracle**, and **MongoDB**).

The following screenshot shows the 3-tier architecture in **Visual Studio** using the **Razor Web App** and **.NET 5** class libraries:

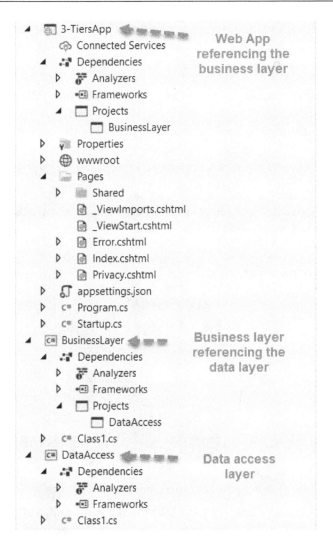

Figure 5.3: 3-tier architecture in Visual Studio

In the preceding example architecture, the **data access layer** is the lowest layer and does not reference any of the other layers. It should contain the **ADO.NET** call or the **EntityFramework** call to manipulate the database tables.

The **business layer** references the **data access layer**. It should contain all the business logic and entities; here, the entities represent the business objects mapped to the database tables. As for the **presentation layer**, this is the **web app** that contains the user interface. It references the **business layer** and doesn't allow direct calls to the **data access layer**. The **web app** can be an MVC app or Razer app, as it is possible to mix these two patterns in order to build a solution.

In the next section, let's get to know the presentation architecture in more detail.

Presentation architecture

One of the major issues we could face in a solution's **User Interface** (**UI**) is the presence of messy code that's difficult to maintain and scale. We have seen this in many web form solutions. This makes the architectural presentation pattern of the utmost importance, as it organizes the source code with a clear separation of responsibilities along with low coupling, which removes any complications and makes the UI code well organized and manageable.

This architecture pattern helps to solve primary UI issues, such as logic that is coupled with the UI, state management, and the synchronization between the UI elements and the business entities.

There are three types of presentation patterns, as shown in the following diagram:

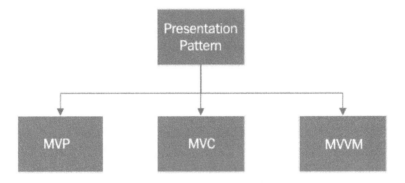

Figure 5.4: The various types of presentation patterns

All of these patterns focus on decoupling the UI from the logic, which allows for clean **HTML** markup. Let's explore these three types in the following sections.

MVC (Model, View, Controller)

The MVC pattern gives you full control over the markup. It is very popular, and Visual Studio has adopted it as the default template for when we want to create a new **ASP.NET** project. It splits the application into three main components:

- **Model**: This encapsulates the business logic and contains the data to display in the view.

- **View**: This displays the content through the UI.

- **Controller**: This handles the user interaction, works with the model for data updates, and, finally, selects a view to render the content.

The following diagram shows the three main components and illustrates which ones reference the others:

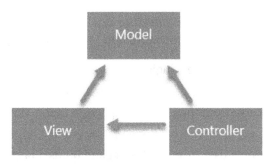

Figure 5.5: The MVC pattern

Here is an example MVC project template using **Visual Studio 2019** and **.NET 5**:

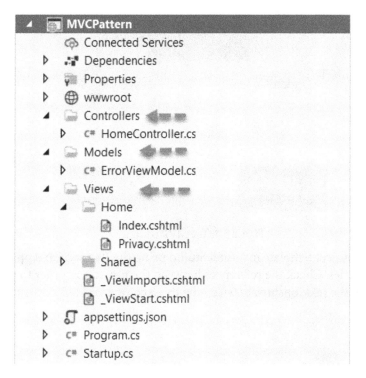

Figure 5.6: A typical MVC project template in Visual Studio

In the preceding example architecture, you can see that the three components are encapsulated within the same project but exist in different folders. You can also create a separate project for each layer and then configure the references.

MVP (Model, View, Presenter)

The MVP pattern is a UI presentation architecture and is considered to be a derivation of the MVC pattern. It separates the architecture into three main components:

- **Model**: This contains the business logic of the solution.

- **View**: This contains the interfaces that enclose the data properties, which we will either send to or receive from the UI. In comparison to the MVC pattern, it doesn't include the UI.

- **Presenter**: This retrieves data from the **Model** and binds it back to the view. It works as an intermediate layer between the **Model** layer and the **View** layer.

The following diagram illustrates the three main components of the MVP pattern:

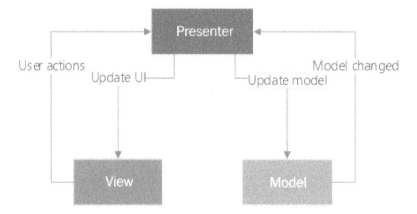

Figure 5.7: The MVP pattern

Here is an MVP project template in Visual Studio using the **Razor Web App** and **.NET 5** class libraries. Check the references between the projects in order to gain an understanding of the relationships between the three components:

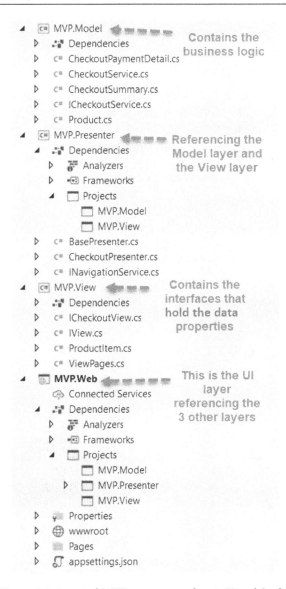

Figure 5.8: A typical MVP project template in Visual Studio

In the preceding example architecture, you can see that the UI is located in the web app, which references the **Model**, the **View**, and the **Presenter** layers. The View layer doesn't reference any other project. It contains the interfaces that are implemented in the Presenter layer. The Model layer contains the business entities and the business logic; it doesn't reference any other project. As for the Presenter layer, this contains the actual implementation of the interfaces defined in the View layer. It references the Model and the View layers because it plays an intermediate role between them.

Let's explore the third type of presentation architecture next.

MVVM (Model, View, ViewModel)

This architecture is also considered an extension of the MVC pattern. It contains three main components, too. It combines the best strengths of MVC and MVP by offering a high level of reusability and scalability. The key concept of this architecture is that it moves the logic out of the controller and into **ViewModel**:

- **Model**: This contains the business rules and the model classes.

- **View**: This contains the UI.

- **ViewModel**: This is an intermediate layer between **View** and **Model**.

The following diagram shows the three components and how they interact with each other:

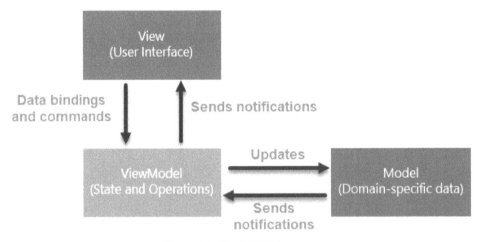

Figure 5.9: The MVVM pattern

Note that the **View** model classes should never use **ASP.NET** state objects, such as `Session`, `ViewBag`, or `TempData`.

Here is an example project template of the MVVM pattern using **Visual Studio**, **MVC**, and **.NET 5**:

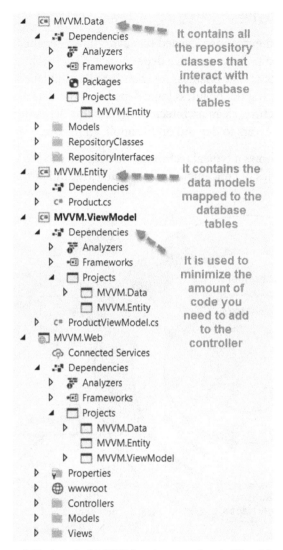

Figure 5.10: A typical MVVM project template in Visual Studio

We can also use the Razor Pages project template as it follows the MVVM pattern. This enables two-way data binding since the code of the model and the controller are attached to the Razor page, which allows for a simple development experience with a separation of concerns.

It is important to understand the difference between these three types of presentation architectures so that you know which pattern you should use for your solution.

Now, let's get to know clean architecture.

Clean architecture

In the n-tier layered architecture, we learned that everything depends on the database layer, which is considered to be a transitive dependency. Clean architecture is considered domain-centric architecture. The business logic and application layers are at the center of the design. Instead of having the business logic depend on the data access layer, as is the case in the n-tier architecture, clean architecture inverts this dependency by forcing the infrastructure and other layers to depend on the application core.

The following diagram shows a typical technique that can be used to visualize this architecture. It uses a series of concentric circles, which are similar to the rings of an onion:

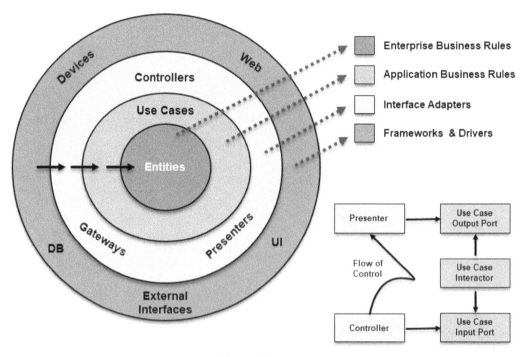

Figure 5.11: Clean architecture onion view

In the preceding diagram, the circles represent the different components of the system, and the application core consists of the entities and the use cases. The key factor influencing this architecture is the dependency rule. It forces the dependency of the components to flow toward the center. The components in the inner circle should never depend on anything in the outer circle.

As a result, if we declare a function or a class in the outer circle, it should not be visible in the inner circle. In the diagram, you can see how the dependencies flow toward the center from the most outer circle to where the use cases and the entities are located.

Let's explore the key components of this architecture:

- The entities represent the business rules, such as objects and related methods.

- The use cases represent the application core, where all the use cases of the system are implemented. It manages the flow of data from and to the entities. This layer is not affected by the changes that might occur in the external layers, such as the database or the UI.

- The interface adapters are the layers that will contain the MVC components, such as the controllers, the views, and the presenters. This layer plays an intermediate role in converting the data coming from the use cases and entities into a suitable format for the external components, such as the database and the controllers. The models in this layer are used as data structures to exchange data between the use cases and the presenters or the views.

- The frameworks and drivers represent the outermost layer of this architecture. This layer contains the database and all of the UI code.

The following diagram shows a horizontal view of this architecture:

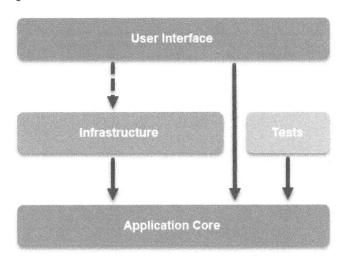

Figure 5.12: Clean architecture horizontal layer view

In this diagram, there are two types of dependencies, which are represented by the style of the arrow. The solid arrow refers to compile-time dependencies, while the dashed arrow represents a runtime-only dependency. The **User Interface** layer deals with the interfaces defined in the application core with no direct access to the implementation in the **Infrastructure** layer. These interfaces are bound to the concrete implementation at runtime through **dependency injection**.

> **Tip:**
>
> Dependency injection is a software design pattern. The main purpose of this pattern is that it allows you to have loosely coupled code that supports architecture patterns, such as clean architecture, to reduce the tight coupling between the layers. It replaces the hardcoded dependencies between the classes by using a builder object to initialize objects and then inject those dependencies at runtime.

The following screenshot shows a clean architecture solution template in **Visual Studio** with **.NET 5** and **Angular 10**:

Figure 5.13: A typical project template using clean architecture

Let's explore each project in this .NET solution:

- The most inner layer is the Domain project. It doesn't reference any other layer, as shown in the following screenshot. On the right-hand side, you can see one of the entity's classes:

```
▷  C# Application                          using System;
▷  Application.IntegrationTests
▷  Application.UnitTests                    namespace CleanArchitecture.Web.Domain.Common
▲  C# Domain                               {
  ▲  Dependencies                             4 references
    ▷  Analyzers                              public abstract class AuditableEntity
    ▷  Frameworks                             {
  ▲  Common                                       3 references
    ▷  C# AuditableEntity.cs                      public DateTime Created { get; set; }
    ▷  C# DomainEvent.cs
    ▷  C# ValueObject.cs                          3 references
  ▲  Entities                                     public string CreatedBy { get; set; }
    ▷  C# TodoItem.cs
    ▷  C# TodoList.cs                             8 references
  ▷  Enums                                        public DateTime? LastModified { get; set; }
  ▲  Events
    ▷  C# TodoItemCompletedEvent.cs               8 references
    ▷  C# TodoItemCreatedEvent.cs                 public string LastModifiedBy { get; set; }
  ▷  Exceptions                                 }
  ▲  ValueObjects                            }
    ▷  C# Colour.cs
```

Figure 5.14: The entities project in clean architecture

- Next, we have the `Application` project. As you can see, it references the `Domain` project. It contains all of the application logic by defining the interfaces that will be implemented in the infrastructure layer:

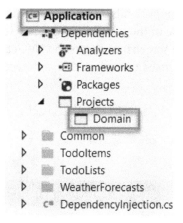

Figure 5.15: The Application project in clean architecture

- Next, we have the `Infrastructure` layer, which contains the concrete implementation of the interfaces defined in the `Application` layer. As per the following screenshot, you can see that it references the `Application` layer:

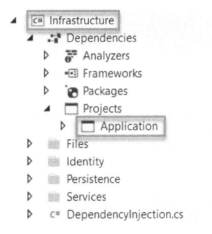

Figure 5.16: The Infrastructure project in clean architecture

- The last layer is `WebUI` (please refer to *Figure 5.13*). This is a single-page application that uses **.NET 5** and **Angular 10**. This layer references the `Application` layer and the `Infrastructure` layer.

Many .NET solution templates support this architecture, which can be found on **GitHub** or **NuGet**. You can download one of the templates from NuGet to get started using the clean architecture. The following screenshot shows the NuGet command line required to install the same template that we used to explore this architecture:

 Clean.Architecture.Solution.Template 1.1.4

Clean Architecture Solution Template for Angular 10 and .NET 5.

.NET CLI

```
> dotnet new --install Clean.Architecture.Solution.Template::1.1.4
```

ⓘ This package contains a .NET Core Template Package you can call from the shell/command line.

Figure 5.17: The command line required to install the clean architecture solution template

In this section, we learned about clean architecture along with its main components. In the next section, we will explore microservices architecture, which is considered to be one of the modern architectures.

Microservices architecture

The microservices architecture allows you to divide the solution into various components. Each component is completely independent of the other components, and it provides a particular service. The following diagram shows the microservices architecture:

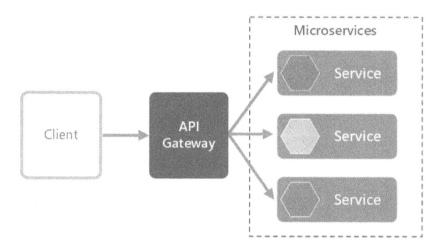

Figure 5.18: The Microservices architecture style

As per the preceding diagram, you can see that the microservices architecture consists of a collection of independent services. Each service is self-contained and should provide a single business capability within a business domain.

Let's examine the characteristics of this architecture pattern:

- **Microservices** are small, autonomous, and loosely coupled services. Each service has its own code base, and it can be developed and maintained by a small team of developers.

- Each **service** should be self-contained and deployed separately. Updating one service won't require you to redeploy the entire solution.

- The services are responsible for having their own data access layer, as each service has a private database.

- The internal implementation of each service is not visible nor accessible by any other service. The communication between the services is achieved through proper APIs.

- The **Client** app has no direct access to the services. Consuming these services is achieved through the API gateway, which forwards the call to the appropriate services.

The following screenshot shows a basic microservice project:

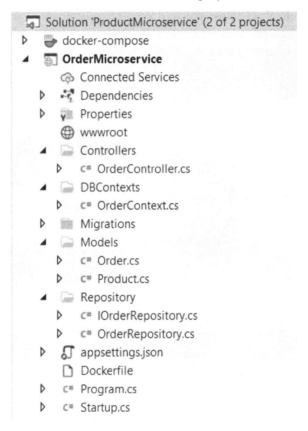

Figure 5.19: An example microservice project

In the preceding screenshot, you can see an order microservice that is consumed by an e-commerce solution. The main components of this microservice project are as follows:

- `Models`: These are the data entities that hold the properties mapped to database fields.

- `OrderContext`: This derives the entity framework, `DBContexts`. It is a bridge between the entity order and the database.

- `OrderRepository`: This holds the CRUD functions, such as `GetOrder`, `CreateOrder`, and `UpdateOrder`. This class should implement the `IOrderRepository` interface.

- `OrderController`: This is a class that is derived from `ControllerBase`. It contains all of the API RESTful actions.

- Docker: This is the container that should simplify the deployment and testing of the microservice by bundling it along with all its dependencies into a single unit. It allows you to run the microservice in an isolated environment.

The web API requests are handled by the OrderController class. The controller will call a function inside the repository that will use DBContexts along with the model to communicate with the database in order to return, add, or edit the requested data.

In the next section, we will explore service-oriented architecture.

Service-oriented architecture

Service-Oriented Architecture (**SOA**) allows you to consume services that are available in the network. Its structure is similar to n-tier architecture; the difference is that the presentation layer can't call the business layer directly, that is, it can only do so through the services. The following diagram shows the SOA in a layered structure:

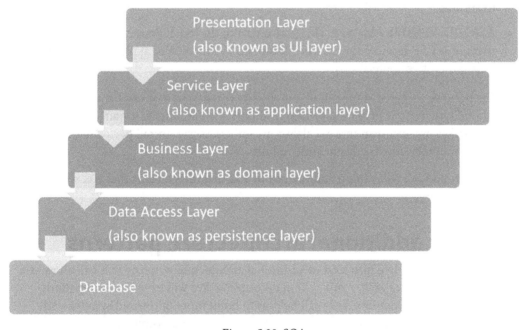

Figure 5.20: SOA

As you can see, the **Service Layer** is an abstraction layer located between the **Presentation Layer** and the **Business Layer**. With the existence of this layer, the **Presentation Layer** doesn't need to communicate directly with the **Business Layer**. In this scenario, you could change the **Business Layer** without affecting the **Presentation Layer**.

What is the difference between microservices architecture and SOA? Well, the answer is the scope. Microservices architecture is a cloud-based architecture; it promotes autonomous services that are self-contained, which target the application scope. While the SOA has an enterprise scope, each service does not need to have an independent database. It can handle multiple business capabilities; this is not the case for microservices, which only handle one single business capability at a time.

Here is a list of .NET technologies that support you in the implementation of services (SOA):

- **.NET Web service**: This is based on the **Web Services Description Language** (**WSDL**) (which is also known as **XML Web Service**). It is a service layer that contains a set of functions that uses a standardized XML messaging system.

- **Windows Communication Foundation** (**WCF**): This is part of the .NET Framework and is used to build service-oriented solutions. By using WCF, you can send any type of data, such as asynchronous messages, from one endpoint to another.

- **ASP.NET RESTful Web API**: This is also part of the .NET Framework. It is used to build HTTP services that can be consumed by any type of application including web and mobile applications.

In this section, we explored some popular architecture patterns that can help us to create a solid foundation for our proposed solution. Adopting an architecture pattern is vital because it makes our platform more scalable and enhances the overall performance of the product. Additionally, it will prevent code redundancy.

In the next section, we will explore another set of architecture patterns that you should know about.

Exploring additional architecture patterns

In this section, we will dive into a set of additional architecture patterns that will allow you to perform high-level scalability and system decoupling. We will examine each pattern to understand how it functions. This will help us to build more optimized systems with reusable modules and an organized structure that allows for extensibility.

The serverless pattern

Serverless architecture promotes cloud platforms and cloud-native code. It is a pattern that allows us to host our solution in a third-party infrastructure. Using this approach, the developers will no longer have to worry about managing the server software and hardware. This pattern allows us to break up our application into small and autonomous functions that can be triggered and scaled individually.

The following diagram illustrates the serverless architecture of a single-page web application:

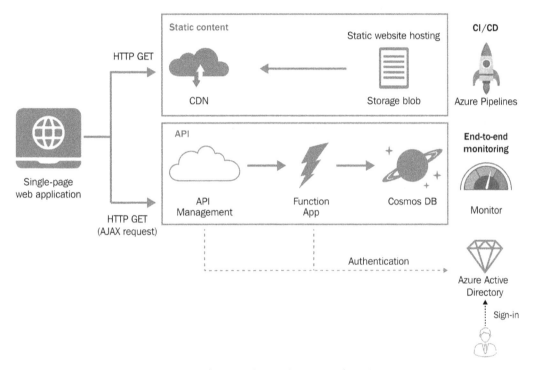

Figure 5.21: The serverless architecture of a web app

We can use the **Azure** serverless infrastructure to implement this pattern. Azure functions include the following:

- **CDN**: This stands for **Content Delivery Network**. It caches the content for a better response time.

- **Azure Blob Storage**: This allows you to store large, unstructured data on **Microsoft**'s data storage platform.

- **Function App**: This is an event-driven model that provides the capabilities to create autonomous functions that are triggered by the client through HTTP requests. The routing of the requests is managed by the API gateway, which is described next.

- **API Management**: This is an API gateway that is located in front of the functions. It allows you to decouple the frontend app from the functions located in the backend. With this API management, we can rewrite the HTTP URLs and manage requests before they reach the concrete functions in the backend. Azure API management is also used to overcome cross-cutting concerns such as the following:

  a. Caching HTTP responses

  b. Monitoring and audit logging HTTP requests

  c. Enabling **Cross-Origin Requests Sharing** (**CORS**), which enables access across domains

  d. Enforcing policies such as checking HTTP requests and applying call rates

  e. Protecting your API by enforcing an authentication mechanism by using **Auth2.0 authorization** with **Azure Active Directory (Azure AD)**

- **Azure Cosmos DB**: This is a **NoSQL** database service provided by Azure to build modern applications.

- **Azure AD**: This is the cloud version of the regular active directory. It is used to authenticate users.

- **Azure Monitor**: This collects performance metrics about the solution and the usage of resources.

- **Azure Pipelines**: This is another service provided by Microsoft Azure. It provides **Continuous Integration** (**CI**) and **Continuous Delivery** (**CD**) services to automatically build, test, and deploy your code to any accessible target.

This pattern is perfect if you wish to implement the microservices architecture or if you want to scale your solution and benefit from pay-as-you-go services.

The client-server pattern

A **client-server pattern** is a network architecture that involves two types of entities: the clients and the server. It is used in scenarios where you have a server playing the role of a service provider and multiple clients playing the role of service consumers. The following diagram describes the logic behind this architecture:

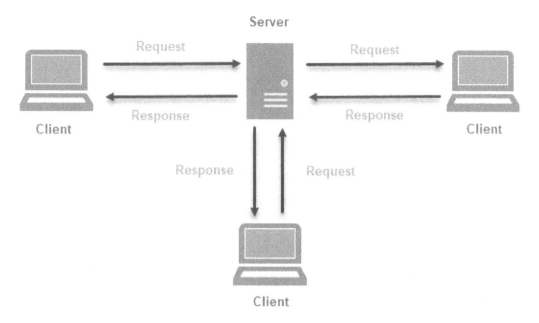

Figure 5.22: Client-server architecture

As you see in the preceding diagram, the **Client** components send an HTTP request over the **TCP/IP** protocol to the server. The request is processed, and the server connects to the database **Server** and then responds back to the **Client**.

The event-driven pattern

Event-driven architecture is another pattern that allows us to highly decouple our applications. It is a pattern that consists of a set of services. Each service works asynchronously and publishes an event when its data is updated. The client components subscribe to the events to receive or send updates.

Let's assume that we are building an e-commerce solution where customers are using coupon codes while submitting an order. The system must ensure that the coupon code is only used once by the same customer. Since the customer information, the number of orders, and the coupon code details are located in different databases, the system cannot simply verify the usage of the coupon code. The solution is to use the event-driven pattern to maintain data consistency across the different services.

The following diagram shows the Azure event-driven architecture:

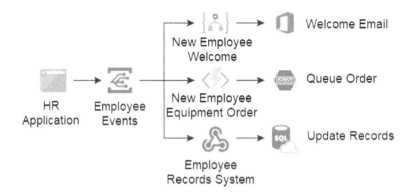

Figure 5.23: Event-driven architecture

In the preceding diagram, pay attention to how the **HR Application** is subscribing to the **Employee Events** and each event has its own logic and database.

The pipe-filter pattern

The **pipe-filter pattern** consists of splitting a complex process into a group of smaller tasks. This approach is expected to improve the performance of our application as well as the reusability and maintainability of each task. A single event triggers a sequence of multiple steps, with each performing a specific task. The following diagram shows an example process that has been implemented using the pipe-filter architecture:

Figure 5.24: The pipe-filter pattern

A good example of this pattern is **Azure Data Factory**, which allows you to create data-driven automated workflows in the cloud for data extraction, analysis and transformation, and loading.

In this section, we explored a set of architectural patterns that you should know about. Each one offers a unique methodology of implementation and delivers key advantages to your solution. In the next section, we will learn how to choose the right architecture pattern when designing a solution.

Choosing the right patterns

In the previous sections, we learned about major architectural patterns. However, you must have noticed that there are a few patterns that we didn't discuss in this chapter. In fact, some patterns will be introduced in the future. So, you will need a way in which to analyze a pattern and decide whether you want to choose it or not. One question that you could be asked is *how do we choose the right pattern for our solution?* Let's dig deep into this matter.

Architecture design is the cornerstone of a solid and successful system. However, there is no one-size-fits-all solution when it comes to choosing the right architecture for your solution. Various perspectives should be taken into account when you want to decide which architecture pattern to use.

Put simply, the main selection criterion for choosing an architecture pattern is based on three factors:

- **Software engineers** (who will work on the project): Software engineers should be familiar with the architecture you are proposing. This is so that they can easily navigate through the solution structure and start implementing new features as per the requirements. Then, at a later stage, when the product is delivered, it should be clear and straightforward to them how to fix any defects and which layers need to be modified. This is a very important factor to consider in order for the team to work efficiently and deliver a successful and stable product.

- **Client**: From the clients' perspective, they are looking for a good quality product while maintaining efficiency. Additionally, they want to make sure that the implemented architecture can support additional features and modifications even after releasing the product. They want their product to be well-architected, scalable, and easy to maintain.

- **Product type**: Note that it is not a good practice to select an architecture pattern just because it is widely popular or because it is trendy. Don't assume that this will deliver a better product. However, an architecture pattern should be selected based on your requirements and the type of solution we want to build. This will allow you to deliver a successful product.

We have explored the three main factors that should affect your selection of the architecture. Furthermore, here is a criterion list with key characteristics that you should consider when selecting an architecture pattern:

- **Agility**: We should consider choosing the architecture that allows for high agility, which helps us to embrace and implement additional features and changes easily.

- **Ease of deployment**: The architecture we choose should allow us to easily deploy the product.

- **Testability**: The architecture should allow for a high testability rate.

- **Performance**: This is an important factor. The architecture should allow for a high-performance rate.

- **Scalability**: The architecture must allow us to scale our system, which means increasing its capacity.

- **Ease of development**: The architecture should be well known by developers to ensure the easy development and implementation of the product. It should allow developers to troubleshoot the system and fix defects when needed.

One of the primary reasons that might cause complete system failure is choosing the wrong architecture pattern. That's why it is important to choose the right architecture pattern for your system, as it will solve various problems that you might face during the several phases of the project life cycle.

Summary

In this chapter, we explored key architecture patterns that are widely used in many solutions along with some modern patterns, such as clean architecture and microservices architecture. We also learned about a set of important architectural patterns that allow you to perform system decoupling and scalability. Finally, in this chapter, we explored the key factors that you need to consider when choosing the right architecture pattern.

In the next chapter, we will dig deep into core architecture considerations, such as the design quality attributes and how to properly plan for system caching, exception handling, and deployment.

6
Architecture Considerations

In the previous chapter, we learned about a set of architectural patterns that you must know to design and build a successful solution. These patterns are essential as they set the path for the development team and address the client concerns related to solution agility, scalability, and performance. Later in the chapter, we had a quick overview of a set of architecture characteristics that affect how you choose the right pattern for your solution.

In this chapter, we will dig deep into the quality factors that should be taken into account, such as reusability, usability, performance, security, development time, and similar quality requirements. Additionally, we will get to know best practices to plan for exception handling, tracing, and deploying.

Here are the topics that we will cover in this chapter:

- Exploring the design and runtime quality attributes of the solution architecture
- Learning how to plan for exception handling, tracing, and deploying
- Caching in web applications

By the end of this chapter, you will have learned what makes an architecture pattern the best fit for the product you are building by exploring design and runtime quality attributes. We will also enrich your knowledge by learning best practices to implement caching to improve performance and overall user experience, logging to track reported errors in a centralized location, and deployment techniques.

Learning about quality attributes

An organized solution architecture sets the right path for your development team and makes it easy to maintain different quality characteristics, which will further enhance the quality of the product in many ways.

Before we start exploring the various quality attributes, let's first understand what a quality attribute is. A **quality attribute** is a property that defines the quality of a system, it is a measurable or testable characteristic of a system that is used to indicate how well the architecture chosen for the system satisfies the requirements of the client. There are two types of quality attributes: qualities that can be measured at design time and others that can be measured at runtime or during execution. The following diagram shows us the various attributes that we will discuss in this chapter along with their respective types:

Figure 6.1: Software architecture quality attributes

Let's explore these quality attributes in the following sections.

Exploring design quality attributes

Business functionalities of the product take the front seat in terms of importance for the development team. We often focus on meeting the functional requirements of the client and later, after releasing the product, we notice some quality deficiency such as the product is difficult to maintain or to scale. Also, we may end up with performance issues or security breaches. In this section, we will explore the design quality attributes that should be addressed during the software architecture phase.

Maintainability

Maintainability is one of the key software quality attributes. It refers to the ability of the architecture to support future changes in the product behavior such as introducing a feature with new business requirements or modifying an existing one.

Repair philosophy also affects the measurement of this attribute, which refers to the time needed to restore the system after a failure. The more our code is coupled and the components are developed with excessive dependencies, the more the product is difficult to maintain. With the existence of this attribute, software engineers started introducing the concept of separation of concerns into architectures, which is supported in microservices architecture, for example.

Improving the maintainability of the product can greatly improve team productivity and lower the cost of adding new features. Here is a list of key techniques for better maintainability:

- Choose an architecture that allows us to separate the responsibilities of the components by having low coupling, which should create well-defined layers in the system and ease changes in the system.

- Use interfaces to maximize the use of plugin modules in the system, which will allow more flexibility and extensibility.

- Provide detailed documentation to explain the object-oriented structure in the system.

Flexibility

Flexibility refers to the ability of the architecture to adapt to varying environments and situations in response to different user and system requirements, which could be hardware changes, software changes, or even changes in the business rules. The less effort you put in to cope with changes, the more flexible it is; the easier it is to reconfigure and deploy the system, the more flexible it is.

A flexible software architecture is able to adapt to changes, so let's check the following key techniques to improve flexibility:

- Consider using business layers to encapsulate the business rules. We can only modify these layers when business rules change.

- Consider using a configurable business workflow engine such as **Microsoft Power Automate**.

- Consider implementing layers in the system to separate the UI from the business logic and the data access functionality.

- Design layers to be consistent and loosely coupled to maximize flexibility and facilitate the replacement and reusability of the components.

Reusability

Reusability is also one of the key software quality attributes. It refers to the degree to which existing components, classes, and functions can be reused to develop new modules, new features, or even new applications. It eliminates the duplication of code and minimizes the time needed to implement new components.

Reusability is an approach that should be applied with careful consideration of the benefits it brings to a system. Here are some key techniques to improve reusability:

- Identify the cross-cutting functionalities between components and implement the common classes and functions that we can reuse across different components to provide capabilities such as validation, logging, tracing, authorization, and authentication.

- Consider exposing the business logic through web services or Web APIs to provide this logic to different systems or platforms, such as web and mobile.

- Use data types and structures that can be accessed through different components.

Integrability

Integrability outlines the way the components are designed to operate together by exchanging information as part of the overall system architecture. It also includes the coding standards and naming conventions in addition to other factors that affect the consistency of the components and makes it easy for the developers to understand the code and maintain it. It also measures the ability of the system to be integrated with other systems.

There are numerous advantages of applying integrability to improve the harmony between the different components of a system. Here are some key techniques to maximize integrability:

- Enforce coding standards that should be predefined and available for the development team and provide detailed documentation for the entire system architecture.

- Consider using web services or gateway layers to integrate with legacy systems.

- Perform code review sessions to ensure guidelines are respected during the implementation.

Testability

Testability is a quality attribute that shows how well a system allows us to create test cases and execute test plans to determine whether the system is working as per the requirements. It allows us to identify faults in the system in an effective manner and based on predefined test cases.

We should find defects, performance issues, and security vulnerabilities sooner as it is less expensive than having the customer find them when the product is released. Let's get to know some key techniques to improve testability:

- Create test cases in **Visual Studio**, then run test plans and check the test results. This is also applicable in **Azure DevOps**.

- Use mock objects in test cases to build different scenarios.

- Let our architecture support modular components to allow detailed testing.

- Create unit testing to test every single functionality in the system.

It is recommended to consistently increase our learning curve and upgrade our skills to ensure that we are able to apply all these design attributes. This will lead to the creation of balanced and highly efficient software solutions and products.

In the next section, we will explore runtime quality attributes.

Understanding runtime quality attributes

Runtime quality attributes are a set of attributes that are measured during the execution of a system in real-life scenarios. They represent a set of features that facilitate the measurement of the performance and security of a software product in addition to other quality constraints.

These quality attributes must be assessed to take actions proactively to ensure they are maintained properly to deliver great products to end users. What follows is an introduction to each runtime quality attribute with some key techniques diving into details that should be considered for improvements.

Performance

Performance is the most important quality attribute for every client. It refers to the responsiveness of the system to perform a specific function in given constraints such as time, accuracy, or memory usage. It includes two metrics, namely, *latency*, which is the time needed to respond to an event triggered in the system, and *throughput*, which is the number of events that can occur in a given time frame.

We all know that there are some products out there that aren't being used because of their performance problems. So, let's get to know some key techniques for improving performance:

- Consider using asynchronous calls.
- Use **Data Transfer Objects** (**DTOs**) to minimize the size of data sent from the server to the frontend client.
- Avoid retrieving data more often than is necessary and use paging when returning data collection.
- Use performance profiling tools, such as **Visual Studio Diagnostic Tools** to identify code that has a large impact on performance.
- Minify frontend assets such as **JavaScript** and **CSS** files.
- Consider using **Azure Functions** to handle long-running requests, as out-of-process functions are beneficial to minimize CPU usage.
- Reduce the size of HTTP responses by using HTTP compression.
- Always use the latest release of ASP.NET Core as it includes many improvements.

Security

Security is an essential part of the system. It refers to the fact that any system should be protected from disclosure and unauthorized attempts to access data. Securing a system starts with implementing proper authentication and authorization mechanisms. In addition, securing the system assets from unauthorized modification is a must. That's why we should always deploy the compiled assemblies and never upload the .NET classes as is.

To secure our system, we must have an in-depth understanding of the environment where we want to deploy the product, what type of access we need to grant users, and what they can access. It is important to know that we need to apply various mechanisms to increase the protection level.

The more we learn about potential threats and take action to avoid them, the more we protect the system. Having the product tested on a regular basis for security vulnerabilities is a must for protection against data breaches that may affect the client's reputation negatively and undermine their brand's integrity. Let's check the following key techniques that should help us improve the overall security of the product:

- Create a periodic task to back up the database and the system assets then store them in a secure location, which will make it possible to recover them quickly when needed.

- Test the restore process to make sure that the backups will work properly.

- Apply solid authentication and authorization processes.

- Never trust user input, always validate data input, and use stored procedures to prevent **SQL injection**.

- Never use string concatenation to create SQL statements.

- Encode passwords saved in the database.

- Do not store sensitive data in hidden fields.

- Implement audit logging functionality to log every single event in the system.

- Consider implementing a clustered server architecture if the system is considered mission-critical for the client.

Reliability

Reliability is the ability of a system to perform all tasks and events triggered by users over time without the need to conduct a repair or modification. The system has a probability of high reliability during the early stage of operation. This probability will start reducing over time. Improving the reliability of a system requires us to identify the most essential user journeys, then analyze them to detect the areas where we can improve. This methodology will allow us to create indicators about the services and functions that matter most to the users.

This quality attribute is critical for the continuity of services delivered by the system. Here are some key techniques to improve the reliability of our system:

- Trace the performance of the most used services in our system to identify poor performance or failures.

- Audit calls to Web APIs and web services.

- Consider implementing a failover plan.

- Consider analyzing customer complaints to troubleshoot and identify the services that should be improved.

Usability

Usability is a quality attribute that assesses the user interface of a system. It shows how easy it is to use the system. If users don't like the design or if they find it difficult to find what they are looking for, they might stop using the system. That's why usability is one of the main factors that will lead users to adopt a system. There are five key factors that constitute the usability attribute:

- **Learnability**: This factor tells us how easy it is for users to perform their tasks the first time they see the system.

- **Efficiency**: This factor specifies how quickly the users can perform their main tasks.

- **Memorability**: This denotes how easy it is to remember the steps to perform main tasks after not using the system for a long time.

- **Errors**: This stipulates how many errors they encounter while performing actions in the system and whether it's easy to report them or to recover and proceed to accomplish the task.

- **Satisfaction**: This indicates how satisfied the users are with the overall design.

Usability concerns should be carefully considered during the earliest design decisions of the system to avoid the disappointment and frustration of end users when the product is released. Here is a list of some key techniques to improve usability:

- Consider maximizing ease-of-use patterns by enforcing accepted UI design standards.

- Simplify user interaction and multi-step functionalities by applying workflows.

- Consider using asynchronous calls to increase user interactivity and to perform background tasks and avoid full post-back calls.

Interoperability

Interoperability is a quality attribute that assesses the ability of the components in our system to cooperate at runtime to perform tasks successfully and efficiently exchange information.

Moreover, interoperability is an attribute of the system that is responsible for its operation and interaction with other systems as well. It is an attribute that cannot be ignored. Let's get to know a few key techniques to increase interoperability:

- Consider using connectors and web services to connect to third-party systems and exchange data.

- Expose functionalities through standard web services or **REST APIs** to exchange data with legacy systems.

- Ensure that our architecture design allows low coupling between components to improve flexibility and reusability.

In this section, we explored the runtime quality attributes that affect the quality of the software product. These attributes should be considered and solved during the implementation and execution of the system to ensure the delivery of great products for our clients. In the next section, we will explore the caching mechanism in ASP.NET Core.

Caching in web applications

Caching is a technique that allows us to store frequently used data in memory. Instead of querying the database multiple times to get the same content, we often use caching to store this content and then retrieve it from the memory the next time we request the same content.

Caching is essential to improve performance in ASP.NET Core and the overall user experience of the product. In ASP.NET Core, there are different techniques to cache data. Here is an overview of these techniques:

- **In-memory caching**: In this technique, the memory of the server is used to store the data.

- **Distributed caching**: This technique is used when our app is deployed to Azure or when it is hosted on a farm environment. The cache is distributed across the servers contributing to this farm.

Let's learn how to implement caching in ASP.NET Core.

Implementing caching in ASP.NET Core

In ASP.NET Core, there are two built-in main interfaces that you can use to start caching the content of mission-critical tasks: `IMemoryCache` and `IDistributedCache`:

- `IMemoryCache`: This is an interface that allows us to apply a local in-memory cache.

- `IDistributedCache`: This is an interface that provides us with a set of methods to manipulate the cache in a distributed environment.

IMemoryCache example

The following code demonstrates an example of using `IMemoryCache` to avoid querying the database multiple times to get the same content:

```
public class NewsService
{
    private const string NewsCacheKey = "news-cache-key";
    private readonly IMemoryCache _cache;
    private readonly IDatabase _db;

    public NewsService(IMemoryCache cache, IDatabase db)
    {
        _cache = cache;
        _db = db;
    }
    public async Task<IEnumerable<NewsItem>> GetNewsList()
    {
        if (_cache.TryGet(NewsCacheKey,
            out IEnumerable<NewsItem> news))
        {
            return news;
        }
        news = await _db.getLatestNews<NewsItem>(...);
        _cache.Set(NewsCacheKey, news,
            new MemoryCacheEntryOptions
            {
    //sliding expiration force the cache to become
        expired after 1 day.
        SlidingExpiration = TimeSpan.FromDays(1)
```

```
        });
        return news;
    }
}
```

In this example, we have a `NewsService` class with a method to get all the updates from the database. Instead of querying the database every time, we want to display the new data. So, we decided to use the `IMemoryCache` interface to benefit from its caching mechanism. In the `GetNewsList` method, we are returning the cached version of the data if available; otherwise, we are querying the database then storing the content in the cache.

IDistributedCache example

This interface should be used when the application is hosted on a web farm or a cloud service. This interface doesn't use the local memory of the server. This cache is shared by multiple web servers. There are various options to implement the `IDistributedCache` interface:

- We can use the SQL Server distributed cache. This cache will be stored in a SQL table. For this option, we need to add the following **NuGet** package: `Microsoft.Extensions.Caching.SqlServer`.

- We can use the **Redis** distributed cache, which is an open source in-memory data store that is often used for a distributed cache. For this option, you need to add the following NuGet package: `Microsoft.Extensions.Caching.StackExchangeRedis`.

Here is an example showing how to use the `IDistributedCache` interface for caching:

```
public class NewsModel : PageModel
{
    private readonly IDistributedCache _cache;
    public NewsModel(IDistributedCache cache)
    {
        _cache = cache;
    }
    public string CachedNewsTime { get; set; }
    public async Task OnGetAsync()
    {
        CachedNewsTime = "Cached Time Expired";
        var encodedCachedNewsTime =
            await _cache.GetAsync("cachedNewsTime");
```

```
        if (encodedCachedNewsTime != null)
        {
            CachedNewsTime = Encoding.UTF8.GetString
                (encodedCachedNewsTime);
        }
    }
    public async Task<IActionResult> ResetCachedTime()
    {
        var currentTimeUTC = DateTime.UtcNow.ToString();
        byte[] encodedCurrentNewsTime = Encoding.UTF8
            .GetBytes(currentNewsTime);
        var options = new DistributedCacheEntryOptions()
            .SetSlidingExpiration(TimeSpan.FromSeconds(60));
        await _cache.SetAsync("cachedNewsTime",
            encodedCurrentNewsTime, options);
        return RedirectToPage();
    }
}
```

In this example, we have created a **Razor** page to display the time and two asynchronous methods: one to get the cached time and the other one is to reset the cache.

In the OnGetAsync method, we get the cached time if available. The ResetCachedTime method is used to set the cache object and define the sliding expiration for 60 seconds, which means the cache will be cleared if it is not used within 60 seconds. Otherwise, the time frame of the cache will be extended for another 60 seconds when it is consumed.

In the preceding two examples, we tried to explain the difference between IMemoryCache and IDistributedCache and how to use them. You can find many Microsoft online forums that provide step-by-step examples on how to configure and implement caching in ASP.NET Core.

In the next section, we will explore the logging and tracing mechanisms in ASP.NET Core.

Unified solution for logging and tracing

.NET logging providers are used to log event messages to track the execution of the application and report all code errors or application exceptions in a centralized location. Tracing is used to track and view diagnostic information about a single flow in the system.

Logging and tracing are essential for .NET teams as every time the application fails, we request information to troubleshoot the issue and resolve it. The unified solution for logging and tracing will give you answers to the following questions:

- *Why did the system fail to complete the action?*

- *When did the error occur?*

- *Which function in the code caused the exception?*

- *What was the status of the data exchanged between the functions?*

For traditional solutions that are hosted on-premises, logging and tracing are performed by the same process that runs the executable of the application. As for modern cloud applications that are built with the microservices pattern, each service runs within a specific process. In this case, the logging and tracing are generated by each microservice process.

The following diagram shows the architecture recommended by Microsoft to implement a unified logging and monitoring system using Azure services:

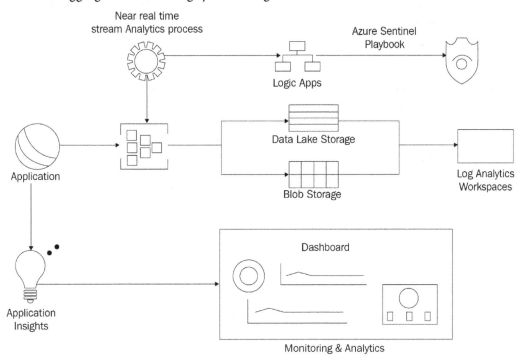

Figure 6.2: Unified logging and monitoring system using Azure services

Let's get to know the main components from the preceding diagram:

- **Event Hubs**: This is a real-time data ingestion service that is fully integrated with all other Azure services. It is used to log all types of events in one centralized hub.

- **Azure Monitor**: This is used to create operational dashboards to help notify .NET teams about any issues and critical malfunctions.

- **Application Insights**: This is part of Azure Monitor, which is used to monitor live Azure services, detect abnormalities in performance, and diagnose and trace malfunctions.

- **Logic Apps**: This is a serverless cloud service that allows you to schedule and organize automated workflows using a user-friendly and easy-to-use visual designer.

- **Blob Storage**: This is cloud storage used for cloud-native workloads to store unstructured data and binary files.

- **Azure Data Lake Storage**: This is a cloud platform that provides secure storage for big data analytics. It provides a set of capabilities required for developers and data scientists to store and analyze big data.

- **Azure Sentinel**: This is a cloud platform that uses built-in AI to log and analyze security information, then report any potential threat or anomalous behavior.

- **Azure Stream Analytics**: This is a serverless cloud engine used to collect and log real-time analytics.

So far, we have learned about the main components. Now let's get an understanding of the logging and tracing mechanism shown in *Figure 6.2*:

1. First, our application hosted on Azure triggers events to **Event Hubs** and **Application Insights** from both a user interface action and a Web API call.

2. **Application Insights** queries log data, traces problems, and monitors the application performance.

3. The **Stream Analytics** platform queries the data in **Event Hubs** to trigger **Logic Apps** workflows and process event messages that are flagged as critical indicators.

4. Then, a **Logic Apps** scheduled process calls a REST endpoint and sends alerts to the .NET teams.

5. **Azure Sentinel** uses **Playbooks**, which are a set of procedures powered by **Azure Logic Apps** to log security alerts or incidents.

6. All logs will then be stored in **Blob Storage** and **Data Lake Storage** for later analysis and troubleshooting.

In the next section, we are going to talk about the high-level deployment steps in Azure.

Planning for deployment and monitoring

In this section, we are going to focus on deploying your ASP.NET application to **Azure App Service**. This doesn't mean that other traditional deployment options are not valid anymore, but we think the future is to host modern apps in the cloud for many reasons, including the capabilities offered by Azure that don't exist in traditional web hosting.

To deploy the ASP.NET Core web app to Azure, we'll need to create a new Azure App Service web app. After the creation of the App Service, we'll be able to deploy our app using **Git** or Visual Studio. To create the App Service, we can use command-line scripts and **Azure Cloud Shell** or you can use the **Azure portal** to create and configure the App Service; both are easy to use.

> **Tip:**
> You can refer to the Microsoft documentation for detailed steps on how to create an App Service. Refer to the *Deploy an app to App Service* section at `https://docs.microsoft.com/en-us/dotnet/architecture/devops-for-aspnet-developers/deploying-to-app-service?view=aspnetcore-5.0`.

After creating the App Service, you can publish the application using Visual Studio. Just right-click the Visual Studio project and then publish it. By default, our app will be deployed to the production environment and we will be able to browse it on the internet.

What if you want to have a staging environment so you can test and approve changes before moving them to production? In this case, you can make use of Azure deployment slots. You can add a new deployment slot that will allow you to swap the app assets along with the configuration settings between the two deployment slots, usually staging and production. You can refer to the Microsoft documentation to create a staging deployment slot (similar steps can be applied to create a production slot): `https://docs.microsoft.com/en-us/azure/app-service/deploy-staging-slots`.

Summary

In this chapter, we explored the design and runtime quality attributes that affect the overall quality of our architecture and as a result, our product. It is important to understand and apply these quality attributes. This will give our product the ability to undergo repair and evolution.

Next, we learned about the impact of caching on the performance of the application and how to enable it using ASP.NET Core interfaces. Later in this chapter, we discussed the logging and tracing mechanism in modern apps, then we explored the deployment capability of Azure App Service.

Remember that our responsibility as solution architects is to get a satisfactory result from the big picture, which consists of the solution architecture as well as the implementation and deployment being done in the correct way – that's what we tried to cover in this chapter.

In the next chapter, we will dig deep into security considerations and will highlight some key techniques to secure your ASP.NET web applications.

7

Securing ASP.NET Web Applications

In the previous chapter, we explored the architecture considerations that should be taken into account when designing and implementing a solution architecture. *Why do we need to learn this?* Because creating an innovative and robust software solution requires us to plan for various aspects and consider different attributes for balancing short-term and long-term product goals and priorities. Paying attention to the quality of attributes, logging, and tracing, along with a proper deployment strategy, will help you deliver a good-quality product that is scalable, maintainable, and secure.

It is exciting for any solution architect to design and build a fancy product; however, this achievement can be ruined if we don't pay attention to the security risks involved. Security is an integral part of any software solution, especially **ASP.NET** web applications. By nature, these applications are exposed to a large number of users, therefore security isn't a luxury in this case and can no longer be an afterthought—it's a necessity.

The **.NET Core** framework provides a set of powerful features and built-in functionalities to secure our applications if we implement and configure them the right way. However, this is not enough, as we still need to apply a set of security measures and write secure code to protect our application from threats and vulnerabilities.

In this chapter, we will cover the following topics:

- Securing ASP.NET Core applications
- Web **application programming interface** (**API**) security recommendations
- Protecting web apps and APIs hosted on Azure

By the end of this chapter, we will have explored a set of security measures, tips, and tricks that will help us build secure ASP.NET web applications. Furthermore, we will get to know some security recommendations to protect our RESTful APIs (where **REST** stands for **REpresentational State Transfer**), along with some tips to securely host our solution on **Azure**.

Most essentially, this chapter provides us with a roadmap to secure our solution. We'll get a deep understanding of how we can incorporate security into our solution architecture, and we'll see what the most important factors are for creating secure software.

Now, let's dig deeper into each of those security measures.

Introducing key security practices

In this section, we will explore key security measures to be taken into consideration while building our ASP.NET web application. There are some **C#** code samples in the following sections that we will use to explain various security vulnerabilities we may face. This code syntax was prepared based on **ASP.NET Core** and **.NET 5**, but the concept is the same even if you have an ASP.NET Web Forms application.

The following is a list of the security measures we will learn about in this section:

- Authentication
- Authorization
- Anti-**cross-site scripting** (**XSS**)
- **Cross-site request forgery** (**CSRF**)
- Cookie stealing
- Overposting
- Preventing open redirection attacks
- Blocking brute-force attacks
- File-upload protection

- Preventing **Structured Query Language (SQL)** injection attacks in ADO.NET and **Entity Framework (EF)**

- General security recommendations

Authentication

Authentication is the process of validating the identity of a user who is trying to access an application. It starts by obtaining the credentials of the user, then validating them against the identity provider such as **Windows Active Directory** that can be on-premises or in the cloud as part of **Microsoft 365** services. The user is considered authenticated if the validation process of the credentials is successful. After authentication, the system should start the authorization process to check the access level of the user and decide which data and resources are accessible for this user. Without knowing who the user is, authorization cannot take place.

There are four different authentication modes in ASP.NET Core that we must know about, as follows:

- **Individual accounts**: This mode is used when we want to make use of the built-in ASP.NET `identity` module. This module will automatically create the authentication and authorization SQL tables, along with the UI that includes the `Register`, `Login`, `LogOut`, and `RegisterConfirmation` views, which will be added to Visual Studio through the scaffolding functionality. The following screenshot shows the SQL tables that will be created when we apply the migrations in the package manager console:

    ▷   ▦ dbo.AspNetRoleClaims
    ▷   ▦ dbo.AspNetRoles
    ▷   ▦ dbo.AspNetUserClaims
    ▷   ▦ dbo.AspNetUserLogins
    ▷   ▦ dbo.AspNetUserRoles
    ▷   ▦ dbo.AspNetUsers

Figure 7.1: ASP.NET identity SQL tables

By using this mode, the unauthenticated users will be automatically redirected to a login page where they can supply their login credentials and submit them back to the server. If the IdP authenticates the request, ASP.NET issues a cookie that contains the ID token of the authenticated user, which will be attached to all subsequent requests in each request header. This means that all subsequent requests are automatically authenticated using the authentication token stored in this cookie.

Here is a Microsoft reference on how to configure this authentication mode:

```
https://docs.microsoft.com/en-us/aspnet/core/security/
authentication/identity?view=aspnetcore-5.0&tabs=visual-
studio
```

- **Microsoft identity platform**: This mode is used if we want to authenticate users against **Azure Active Directory**. We will have to register our app with the Azure Active Directory, then configure our ASP.NET Core project. Here is a screenshot showing the settings that we need to change in the appSettings.json file; we can get these settings from the Azure Active Directory after we register the app:

```
"AzureAd": {
  "Instance": "https://login.microsoftonline.com/",
  "Domain": "qualified.domain.name",
  "TenantId": "22222222-2222-2222-2222-222222222222",
  "ClientId": "11111111-1111-1111-11111111111111111",
  "CallbackPath": "/signin-oidc"
},
```

Figure 7.2 – Configuration in the appsettings.json file

As shown in the preceding screenshot, we first need to set the Domain name that we are using in the application. Then, we need to set TenantId and ClientId that we get from Azure when we register the application. As for CallbackPath, this is the **Uniform Resource Locator** (**URL**) where we want to redirect users after successful login.

The following diagram shows how the authentication with the Microsoft identity platform works:

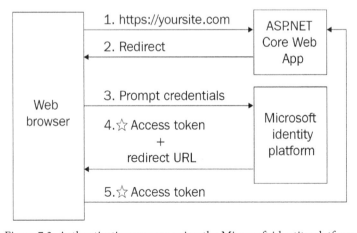

Figure 7.3: Authentication process using the Microsoft identity platform

As you can see, unauthenticated users will be redirected to the Windows login page where they are prompted to provide their credentials, and then an access token is created if the credentials are valid. After that, the user gets redirected to the landing page or redirects the URL specified in the **HTTP** response returned by the identity provider.

Here is a Microsoft reference on how to configure the Microsoft identity platform:

```
https://docs.microsoft.com/en-us/azure/active-directory/
develop/quickstart-v2-aspnet-core-webapp
```

- **Windows**: This is also known as **Negotiate**, **Kerberos**, or **New Technology LAN Manager** (**NTLM**) authentication. This authentication mode is best suited for apps running in intranet environments under the same Windows domain. It can be configured for apps hosted with **Internet Information Services** (**IIS**) or **Kestrel** while the server runs on a corporate network using Active Directory domain identities. This authentication process relies on the operating system to get the ID of the user and confirm the authentication.

Here is a Microsoft reference on how to configure Windows authentication:

```
https://docs.microsoft.com/en-us/aspnet/core/
security/authentication/windowsauth?view=aspnetcore-
5.0&tabs=visual-studio
```

- **None**: When we choose this mode, it means the identity of users is not needed. This type of mode is used in two cases—either when our application is public and anyone is allowed to access it or when we want to build our own custom authentication module.

Here are a few tips to consider when we implement a custom authentication process:

- Enforce the user to use a complex password and hash it before storing it in the users' table.

- Never store a password in a hidden field or in any state management object.

- Consider encrypting the password input using a client-side library before submitting it to the server along with the request header and body. On the server, when you receive the password you will need to decrypt it, hash it, and then compare it to the hashed password in the database. If they are equal, then the user is considered authenticated.

- If you are using sessions, make sure to clear them on logout and modify the session ID, and on login generate a new session ID.

- Consider implementing **two-factor authentication (2FA)**.

- Never grant any user db_owner access to our SQL database, including the user used in the connection string.

Authorization

Authorization is the process of deciding whether a user ID should be granted access to a specific resource in an application. Usually, authorization starts immediately after authentication, and there are different types of resource authorizations that can be given to a user, listed as follows:

- **URL authorization**: This is performed to selectively grant users and roles access to particular URLs in the application.

- **File authorization**: This process is used to protect the assets of an application and prevent unauthorized users from browsing the directories.

- **UI authorization**: This is also known as **UI trimming**. This process is performed to selectively allow or deny access to arbitrary parts of a page for specific users or roles. The section will be completely removed from the page if a user has no access to it.

It's quite easy to apply authorization in **model-view-controller** (**MVC**) by adding the [Authorize] attribute to the controller class or to actions that are not anonymous. Here is an example of this:

```
[Authorize(Users = "john,tim")]
public IActionResult EditContent()
{
  return View();
}
```

If you allow anonymous access to a particular action within a controller class that has the [Authorize] attribute on top of it, you need to use [AllowAnonymous] on top of the action. You can use the [Authorize] attribute to grant access to roles and not only users.

Anti-XSS

XSS is considered the number-one security vulnerability on the web and, unfortunately, a large number of web developers are not familiar with the risks of this vulnerability. XSS is a type of injection attack in which an attacker tries to execute malicious client-side scripts in the web browser of another end user.

There are two scenarios in terms of XSS attacks: the first one is called **passive injection**, where an attacker inputs a vulnerable script in an input field that will be stored in the database and will be displayed on the page when users access it. The second one is called **active injection**, where a user enters a vulnerable script into an input that will be displayed immediately onscreen.

Let's explore these two scenarios with some examples in the following sections.

Passive injection

This type of XSS attack occurs when the web page accepts unsanitized text input that can be later displayed to a victim who is accessing this page. Suppose we have an online blog post that allows users to post comments and interact with each other.

If the input field, where we should specify our comment, is accepting the text as is without validation or sanitization, then the attacker will inject a client script in the comment field, which will be triggered whenever a user is accessing this blog post. Here is an example of a comment that contains malicious input:

```
This is a nice post<script>window.alert('This is an unsecure
website')</script>
```

In this example, the comment contains **JavaScript** code that will trigger an alert with a nasty message. This message will be stored in the comments table, and every time a user tries to access the page, the script will be triggered and the message will be displayed to the end user, which is very annoying.

The attacker can inject JavaScript code to manipulate the **HTML** code of the page, such as in this example:

```
This is a nice post<script src="http://hackersite.xxx/
badscript.js"></script>
```

In the preceding example, you will notice that the attacker injected a client-side library that can manipulate the HTML code of your page and display different content.

Active injection

This type of XSS attack occurs when the user input is immediately displayed on the web page and is not saved on the server. Suppose we have a web page that is reading metadata from the query string of the URL, and it shows a welcome message when we access the page.

In this case, an attacker can manipulate the query string and pass the following input script:

```
johnsmith\x3cscript\x3e%20alert(\x27XSS attack! weak security\
x27)%20\x3c/script\x3e
```

This will display an `XSS attack! weak security` alert message on the web page.

Let's check the following recommendations to help protect your application against XSS attacks:

- Don't trust any user input, even if the user is authenticated. You should always validate the input provided by the end users. Moreover, you should encode query strings and escape single quotes before storing the text in a database or displaying it on a web page.

- Ensure the URL query string is encoded, and always validate the value in the query string before using it.

- Perform content sanitization before you store untrusted content in your database. **HTML sanitization** is the process of checking content that is dynamic and only preserving tags that match with the whitelist.

- You should always use `@Html.Raw` to render untrusted content.

- You can encode untrusted data before displaying it in your HTML code. This way, you make sure no one can inject an input with a script code because the encoding mechanism will convert < to `&lt;`, which will be treated as regular text.

- Make sure to set the `HttpOnly` flag to protect our cookies from being accessible through client-side code.

Cross-Site Request Forgery (CSRF)

CSRF (also known by the acronym **XSRF**, and pronounced *sea-surf* or *c-surf*) is a type of attack that is performed by a malicious website that enforces a trusted but vulnerable site to perform an undesirable action when the user is still authenticated.

A CSRF attack is possible because browser requests include cookies that encapsulate the authentication tokens. In this case, the attacker is taking advantage of the authentication cookie to fool the trusted website, which cannot distinguish between legitimate requests and forged requests, by executing a malicious request using the authentication cookie from the trusted website. This type of attack is also known as a **one-click attack** or **session riding**.

The easiest way to perform a CSRF attack is by attracting the attention of users to a malicious website by sending millions of phishing emails claiming that users won a big amount of money or a trip to Las Vegas. Usually, there is a link included in the email that will take us to the malicious website, and to collect our prize the malicious website would ask us to click a fancy button.

Of course, users would not hesitate to do so, for different reasons. One such reason is that they don't know the risks or the consequences of clicking the button. Once the button is clicked, the malicious website sends the nasty request to the trusted website while attaching an authentication cookie with the request. If the vulnerable website is not taking precautions such as validating the incoming request (as in this case), the attack will, unfortunately, be successful. Here is a diagram showing a CSRF attack:

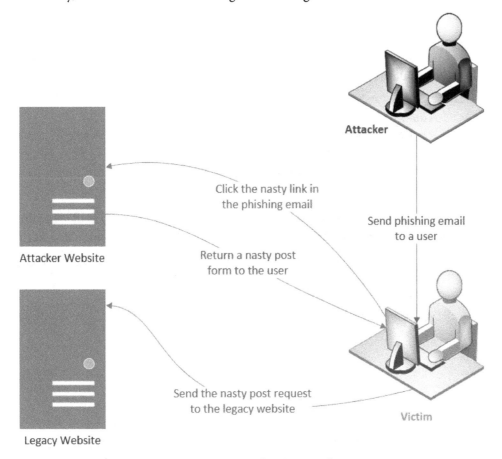

Figure 7.4: Steps of a CSRF attack

Here are some recommendations that should be considered to prevent CSRF attacks:

- Generate a user-specific CSRF token and store it in a hidden field. This token should be submitted with every request, and it should be validated on the server on all POST, PUT, and DELETE requests. The token should be regenerated on every request to prevent attackers from simulating this token and fooling the validation process on the server side. In MVC, we can use the following code to generate an anti-forgery token:

```
@using (Html.BeginForm("Create", "Product"))
{
@Html.AntiForgeryToken()
  //Here we put our form fields along with the submit
    button
}
```

This will output the following HTML code:

```
<form asp-controller="Product"
    asp-action="Create" method="post">
    <input name="__RequestVerificationToken"
        type="hidden" value="sK0JeZQad..AhEYoo1" />
  <!-- rest of form goes here -->
</form>
```

As you can see, the "__RequestVerificationToken" hidden field is holding the CSRF token. Here is an example, showing us how to force the post action in the controller to validate the token before executing the core functionality of the action:

```
[HttpPost]
[ValidateAntiForgeryToken]
public async Task<IActionResult>
    Create (ProductViewModel newProduct)
{
<!-- rest of the action code goes here -->
}
```

Notice the attribute on top of the action to validate the anti-forgery token. If the token is not valid, the request will be rejected/canceled.

- Consider checking the referer header of incoming requests, which should be referencing the same domain of the trusted site. This will prevent or cancel requests submitted from a different domain.

Cookie stealing

Cookies are an essential part of a website because they usually hold the session details of the logged-in user. A **cookie** is an object that is transmitted back and forth between the client browser and the server. So, instead of authenticating the user on every request, the authentication token or claims can be stored in the cookie, and it will be used to identify the user after login. Without cookies, the user will need to log in again on every request.

It is important to secure cookies if you are using them in your application. **Cookie stealing** (also known as **session hijacking**) is a type of attack that allows hackers to steal the cookie of a logged-in user, then impersonate that user and start sending requests on their behalf. In this case, the server is fooled because the request sent by the attacker looks authentic since it contains a valid authentication cookie.

To prevent cookie stealing, we must apply the following recommendations:

- Use a **Secure Sockets Layer** (**SSL**) certificate and only allow **HTTP Secure** (**HTTPS**) requests to encrypt all requests passed between the user and the server.

- Consider applying `secure` and `HttpOnly` flags in the `web.config` file to protect the cookie and to ensure that it is only sent over an SSL connection.

- Regenerate the session ID immediately after login.

- Consider clearing the authentication cookie on logout.

Overposting

Model binding in ASP.NET handles the mapping of data between incoming requests and the .NET application model. It is a powerful feature that simplifies the process of populating the model properties with the user input data, based on a naming convention. However, this may cause another security breach by allowing an attacker to populate some properties in the model that are not presented in the form. This type of attack is called an **overposting** or **mass assignment**.

Let's check the following example to understand the overposting vulnerability. Assume we have a user model that we are using to register a new user in our application:

```
public class User
{
    public int ID { get; set; }
    public string FirstName { get; set; }
    public string LastName { get; set; }
    public string Email { get; set; }
```

```
    public string Password { get; set; }
    public bool IsAdmin { get; set; }
}
```

This is pretty much a simple user model. You will notice that there is a property named
IsAdmin in the model—this is used to specify if the user has an administration access
level on the entire application. The **CSHTML** view should not include this property
because we don't want end users to decide their access level. The view should look like this:

```
@model User

@{
    ViewData["Title"] = "Register";
}

<h1>Register</h1>
<div class="row">
    <div class="col-md-4">
        <form asp-action="Register">
            <div asp-validation-summary="ModelOnly" class="text-danger"></div>
            <div class="form-group">
                <label asp-for="FirstName" class="control-label"></label>
                <input asp-for="FirstName" class="form-control" />
                <span asp-validation-for="FirstName" class="text-danger"></span>
            </div>
            <div class="form-group">
                <label asp-for="LastName" class="control-label"></label>
                <input asp-for="LastName" class="form-control" />
                <span asp-validation-for="LastName" class="text-danger"></span>
            </div>
            <div class="form-group">
                <label asp-for="Email" class="control-label"></label>
                <input asp-for="Email" class="form-control" />
                <span asp-validation-for="Email" class="text-danger"></span>
            </div>
            <div class="form-group">
                <label asp-for="Password" class="control-label"></label>
                <input asp-for="Password" class="form-control" />
                <span asp-validation-for="Password" class="text-danger"></span>
            </div>
            <div class="form-group">
                <input type="submit" value="Create" class="btn btn-primary" />
            </div>
        </form>
    </div>
</div>
```

Figure 7.5 – User registration sample form

146

When the form is submitted, it will produce the following HTTP POST request:

```
Request URL:http://TheWebsiteUrl/register
Request Method:POST
Status Code:200 OK
firstname:John
lastname:miller
email:john@xxx.com
password:encryptedPassword
. . .
```

However, when using a debugging proxy-server tool, we can modify this HTTP request and include additional values and properties in the request. In this case, the attacker would include IsAdmin:True in the POST request. As a result, the user will be registered in the system with admin privileges.

How can we prevent this kind of attack? Well, there are a few solutions to prevent this vulnerability, such as the following ones:

- **Matching incoming parameters**: Instead of using the full model as an input parameter for the MVC action, just declare the fields that we need to pass to register the user. So, instead of public IActionResult Register(User model), we can use public IActionResult Register(string firstName, string lastName, string email, string password), and in the implementation of the action, we populate the model using the fields passed in the action's parameters. In this case, any additional property that is added by the attacker will be ignored.

- **Using a view model**: Again, instead of using the full model as an input, just create a new custom view model and call it RegisterUserViewModel. In this new model, we only add the properties needed for registration, so the action will become public IActionResult Register(RegisterUserViewModel model). I like this option and I usually apply it as a common practice.

- **Whitelist parameters**: We can use a BindAttribute class on the method parameters and just include (whitelist) the properties we want to allow for binding. So, the action should look like this: public IActionResult Register(([Bind("FirstName,LastName,Email,Password") User model).

As a good practice, we must not use our database entities directly as models in the MVC views and actions.

Preventing open redirection attacks

Let's first understand what an open redirection attack is. If you have logic in your web application that redirects users to a URL that is specified in the query string or via the HTTP request's parameters, this can potentially be tampered with to redirect users to a malicious URL, to steal their credentials.

Assume an attacker sent an email with a redirect link such as this:

```
http://www.yourtrustedwebsite.com?ReturnUrl=www.fakedomain.com/
login
```

Usually, some users won't look at the query string, and others won't even check the domain in the first part of the URL. When they click this URL, they will get redirected to a login page provided by the malicious website. This login page is very similar in terms of look and feel to the original login page in the trusted website. In this case, users will provide their credentials, assuming they log in normally.

However, the attacker will steal the credentials and redirect them to the original login page in the trusted website. Users will feel as though they provided the wrong username or password, so they will provide these again and continue what they wanted to do on the trusted website.

In this way, the attacker steals the user credentials without the victim ever knowing about it. This type of attack is called an **open redirection attack**. Now, let's get to the part about how to prevent it.

When using such redirection logic in your web applications, treat all users as untrustworthy. Therefore, we need to make sure to only redirect to local URLs within our application or make use of a new method available in ASP.NET, named `LocalRedirect`. This is used to redirect to a local URL within the app itself, which means it validates the URL before triggering the redirection, and if the URL is not local, the method will throw an exception.

Also, there is a method to validate whether a URL is local or not—you can make use of `Url.IsLocalUrl(..)`. This method will return a Boolean to indicate whether the URL is local or not.

Blocking brute-force attacks

In cryptography, a **brute-force attack** (also known as an **exhaustive search**) involves an attacker attempting to guess a password by thoroughly trying every possible combination of letters, numbers, and symbols until discovering the correct combination that works. In many cases, the attacker will use a bot tool to perform an automatic attack and predict the password. To prevent this type of attack, we can apply the following steps:

- Lock the user account after a specific number of failed login attempts.

- Implement a **Completely Automated Public Turing test to tell Computers and Humans Apart** (**CAPTCHA**) on the login page.

- Consider allowing logins from specific **Internet Protocol** (**IP**) addresses and restrict these from all other IP addresses.

- Enforce complex passwords.

- Consider enabling 2FA.

- Block the attackers' IP addresses, but this is not a guaranteed solution because the attackers can change the IP addresses from which they are performing the attacks.

- Consider using a proper username and avoid using `admin`, `administrator`, or any other easy-to-guess usernames.

File-upload protection

A file upload allows users to upload files while submitting a form. A career form is a simple example of the usage of a file upload, where users need to attach their resume when applying for a job position. Attackers can make use of the file upload and try to upload malicious files to the server. Here are a few security steps that should reduce the likelihood of using a file upload to perform a successful attack:

- Disable the execute permissions on the folder where you are storing the uploaded files.

- Make sure to use a whitelist to only allow approved file extensions.

- Enable client-side validation to check the file extension before uploading it to the server.

- Check the size of the uploaded file and restrict the uploading of large files that exceed the size limit.

- Make sure to check the header of the uploaded file, using the server-side code in .NET, to prevent the upload of malicious files.

- Encode the filename, especially if you are displaying the filename in the HTML code.

Preventing SQL injection in ADO.NET and Entity Framework

A **SQL injection** is a vulnerability that enables an attacker to bypass the security measures taken in an application to execute malicious SQL commands directly in the application's database.

With these SQL commands, attackers can query the data of other users. They can also modify data and even delete some tables or the entire database, which can cause a major loss to the client business, especially if there is no proper backup process in place.

Furthermore, they can escalate an attack to compromise the entire SQL server. A SQL injection attack is one of the most dangerous attacks we can face because it affects the entire database and possibly all databases hosted on the same server. Let's get to know how to prevent this type of attack, as follows:

- Check for malicious input data by enforcing constraints, validating the type and format of the data, and enforcing sanitization.

- Consider using parameterized SQL stored procedures for data access and avoid using text concatenation with inline SQL statements.

- Never grant administrative privileges to SQL users that are used in the data access layer—the read/write permissions should be enough.

- Avoid disclosing the details of database errors that may occur in the application. Actual errors should be logged properly in a centralized location, and the end user should be redirected to a custom error page with no technical details.

- Encrypt the SQL connection in the `web.config` file to secure connectivity with the database.

- SQL injection vulnerabilities are applicable in **NoSQL** databases such as **Azure Cosmos DB** and **MongoDB**, and therefore all the previous recommendations are also applicable in this case as well.

General security recommendations

In the previous sections, we learned about the major security vulnerabilities. In this section, we will highlight some security recommendations that will increase the security level of the solution, as follows:

- Consider enabling audit trails, logging, and tracing to monitor all events and incoming requests.

- Always upgrade the .NET version used in your solution by installing .NET patches to benefit from the security enhancements released by the Microsoft team.

- Consider encrypting passwords before sending them to the server to avoid sniffing attacks.

- One of the common security steps is to enable the following response headers:

  a. `Content-Security-Policy`: This allows us to specify a source whitelist of content that can be loaded onto the website. It helps to prevent XSS, **clickjacking**, and other code-injection attacks.

  b. `X-Content-Type-Options`: This helps in preventing **Multipurpose Internet Mail Extensions-sniffing** (**MIME-sniffing**) attacks.

  c. `X-XSS-Protection`: This enables the XSS filter.

- Block **cross-frame scripting** (**XFS**) attacks by enabling the `X-Frame-Options` response header.

- Prevent disclosing sensitive data related to the hosting server and .NET Framework by removing the following response headers:

  a. `Server`: This header specifies the web server version (IIS version).

  b. `X-Powered-By`: This header indicates that the website is powered by ASP.NET.

  c. `X-AspNet-Version`: This header specifies the version of ASP.NET used.

- Avoid using third-party components and libraries with known vulnerabilities.

- Consider updating **NuGet** packages periodically to make use of the latest fixes and enhancements.

- If you are hosting your app with IIS, make sure to encrypt the connection string because it contains the credentials of the user who can access the database. If you are hosting your app with Azure App Service, consider storing the connection string in the Azure application settings instead of the `web.config` file.

In this section, we explored a set of key security practices to help secure our ASP.NET web applications against malicious attacks. Once each of these practices is applied, it will add a security layer to the application. The objective is to highlight various areas that a solution architect should focus on while designing a robust web solution.

In the next section, we will learn how to secure a web API with a set of security recommendations.

Web API security recommendations

With an increasing demand to build modern web and mobile apps, web APIs have become essential to empower these applications, with an easy way to communicate with the data access layer. This should be accompanied by proper security measures to protect web APIs. In addition to the security recommendations we discussed in the previous section, here are some essential tips to secure your web API:

- Consider using the latest **Transport Layer Security** (**TLS**) version to encrypt communication between the app and the server.

- Authenticate users who are trying to consume the RESTful API.

- Act like a stalker by enabling audit logs, tracing, and logging to monitoring all events.

- Consider protecting your API by applying throttling and quotas, such as limiting the number of messages per a specific time. This is important to control the bandwidth of the server according to the available capacity.

- Always validate the **JavaScript Object Notation** (**JSON**) data input to avoid SQL injection.

- Enable proper firewall configuration on the server where you host the web API.

- Consider having an API gateway, which is a middleware layer that sits between the client application and the RESTful API. This helps you to secure, control, and monitor the traffic to the RESTful API.

- Prevent a **distributed denial-of-service** attack (also known as a **DDoS** attack), which sends a large number of useless requests to overwhelm the memory and capacity of the hosting server by flooding it with concurrent connections. You can prevent DDoS attacks in IIS by enabling the dynamic IP restrictions extension that can block incoming requests from certain IP addresses based on the number of concurrent requests. If the application is hosted in Azure, then we can enable **Azure DDoS protection**.

- Consider enforcing a timestamp in every request by adding it to the request header. This timestamp should be validated on the server to only accept requests if their timestamp is within a particular timeframe. This approach can help you protect the web API against brute-force attacks (explained in the *Blocking brute-force attacks* section) and replay attacks that allow attackers to maliciously complete duplicate requests.

In this section, we discussed a set of security recommendations that should be applied to secure an ASP.NET web API.

Protecting web apps and APIs hosted on Azure

In this section, we will highlight some security recommendations to bear in mind if you are hosting your web application or your web API on Azure, as follows:

- Consider enabling **Azure Defender** to protect your app service.

- Always run the integrated vulnerability assessment scanner available in Azure Defender for SQL servers to extend the protection of SQL servers along with stored databases.

- You can keep your app service up to date by using the latest versions of supported platforms, frameworks, and protocols.

- Disable anonymous access to the blob storage to protect uploaded files. You can enable anonymous access to specific folders if needed.

- Enforce the usage of the **SSL/TLS** protocol to provide a secure connection.

- Always use **File Transfer Protocol Secure** (**FTPS**) instead of the regular **FTP** to deploy your files and disable the FTP protocol if you are not using it.

- Consider using environment variables to store your database credentials, API tokens, and any application settings.

- Consider using a cloud **Windows Application Firewall** (**WAF**), which can help to protect your web applications from malicious attacks and common web vulnerabilities such as SQL injection and XSS.

Summary

In this chapter, we learned that security is an essential part of a web solution. We outlined the fundamental security measures and techniques to help in understanding the possible security vulnerabilities that will allow us to protect an ASP.NET web application against malicious attacks.

Furthermore, we highlighted some key guidelines to secure our RESTful API. Later, in this chapter, we explored some tips to secure our app that can be hosted on Azure. These security practices allow us to build robust yet secure ASP.NET apps.

In addition to the recommendations shared in this chapter, I strongly recommend you keep updating your knowledge about the security features in ASP.NET by reading the online official documentation shared by the Microsoft .NET team. Here is the link to the documentation: `https://docs.microsoft.com/en-us/aspnet/core/security/?view=aspnetcore-5.0`.

In the next chapter, we will explore the different types of testing that we may need to conduct before releasing our solution.

8

Testing in Solution Architecture

In the previous chapter, we learned about how to secure an **ASP.NET** web solution. We also highlighted some key security recommendations to protect our web **application programming interface** (**API**), along with security best practices when it comes to hosting on **Azure**.

In this chapter, we will become familiar with the most common testing types you need to know, and when to use them.

In this chapter, we will cover the following topics:

- Highlighting key testing principles
- Learning about the main types of software testing
- Exploring testing in Azure

By the end of this chapter, you will have learned about the various types of software testing that we can apply to test our software solution, with the aim of finding errors and then fixing them. We will also learn how to check whether the software works properly and whether it meets the requirements defined during the early stages of a project. We will also explore the testing mechanism offered by **Azure DevOps**.

Moving on to the next section, let's take a look at the key principles of software testing.

Highlighting key testing principles

The main objective of conducting software testing is to eliminate possible bugs and to enhance the overall quality of the software in terms of many aspects, such as performance, **user experience** (**UX**), and security.

But before starting any testing activities, there must be some guidelines or principles in place to make sure that the outcome of these activities is aligned with the main objectives of testing. Here, in this section, we will highlight some of the key principles of software testing that we need to consider in our software solution, as follows:

- All test cases should be prepared based on customer requirements; otherwise, we will be testing against the wrong requirements. Each feature or function in a system should be tested with one or multiple test cases.

- Some types of software testing such as **performance testing** and **acceptance testing** should be performed by **subject-matter experts** (**SMEs**) such as **quality assurance** (**QA**) engineers or senior developers.

- Plan to start testing the basic functionalities first, then extend to testing the advanced features.

- It is recommended to start testing at the early stages of a project as, in this case, the cost of fixing defects is way less than when testing during later stages of the project.

- **Defect clustering** is based on the **Pareto** principle, which states that 80% of defects are caused by 20% of the system features. This means that during testing, a large number of defects detected are related to a small number of features.

- It is not recommended to repeat the same test cases over and over because, after a certain time, we won't find any new defects. The best practice is to adjust the test cases, with the aim of finding new defects.

- Testing is context-dependent, which means we need to apply specific methodologies and techniques based on the context of the system we are testing. For instance, testing a **content management system** (**CMS**) is different from testing an **iOS** e-commerce app.

Let's start exploring the various types of testing.

Exploring the main types of software testing

One of the major reasons for failure in software projects is a lack of product quality. Software testing is an integral part of the project life cycle, helping to ensure that a product is error-/defect-free and, in the same way, verifying the functionalities implemented to make sure they match the requirements defined with the client. There are two main categories of software testing, outlined as follows:

- **Functional testing**: This is used to validate each feature and function of the system to verify all functionalities.

- **Non-functional testing**: This is used to validate non-functional aspects of the system, such as performance, usability, and compliance.

Here is a diagram showing the different types of testing we will discuss in this chapter:

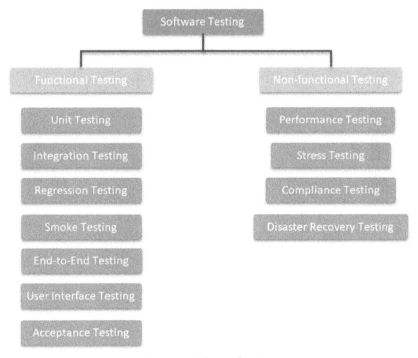

Figure 8.1: Types of testing

We will look at all the different types of testing shown in the preceding diagram in the upcoming sections.

Unit testing

Unit testing is a type of testing performed to test every individual function or module of a system. Usually, it is performed by .NET developers who are working on a product because it requires some coding skills. That's why it is considered a low-level type of testing since it is targeting the behavior of the code only.

Here is a diagram showing the unit-test level as an integral part of the entire testing life cycle:

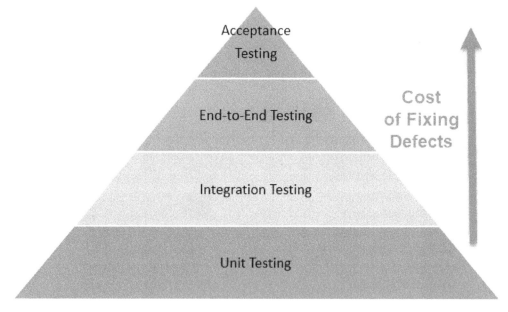

Figure 8.2: Unit-test level in the testing life cycle

In the preceding diagram, unit testing represents the first type of testing that should be conducted before starting any other testing activity, as the cost of fixing defects becomes higher at later levels of testing.

Here are some benefits of unit testing:

- Unit tests allow us to fix defects at the early stages of the development cycle. This will save time and costs to fix the same defects later on during the acceptance-testing stage.

- It helps to document the source code.

- It allows the developers to refactor the code and reuse available functions to eliminate any repetition in the API.

- Unit testing is essential for testing dependencies if we are making changes to the API.

- It helps reduce code complexity.

For more details on automating unit tests, see the list of testing tools recommended by Microsoft: `https://docs.microsoft.com/en-us/dotnet/core/testing/#testing-tools`.

Integration testing

Integration testing is intended to test two or more modules of a solution to verify whether they work well together. For example, it can involve testing the behavior of the integration between the system we are building and **Azure Active Directory**, to verify the authentication mechanism.

Another example of this type of testing is when we need to verify the interaction between our system and the database layer. Integration testing should be performed after completing the development of two modules that are subject to the testing we are conducting.

In the following diagram, we are showing that integration testing should target only the integration part between **Module A** and **Module B**:

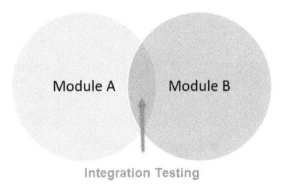

Figure 8.3: Integration testing for two modules

Here are some benefits of integration testing:

- Integration testing helps to ensure that the integrated modules are working properly as expected.

- It allows for simulating the transition between various modules in the system.

- It also helps to detect errors that may occur in the interaction of the modules.

Regression testing

It is normal to test new changes that we perform on a system, such as modifying an existing feature or adding a new one. However, this is not enough, because in most cases, the code we change or add will have a direct or indirect impact on other functionalities, and probably on other features in the system too. This is why we need to conduct **regression testing** to make sure that the new code didn't cause any new defects.

In the following diagram, we are showing the three main steps of regression testing:

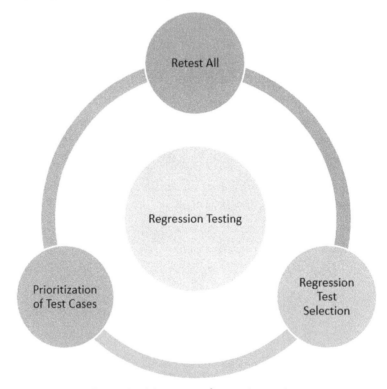

Figure 8.4: Main steps of regression testing

Here are some benefits of regression testing:

- Regression testing ensures that existing features remain untouched in case of a change to a module or code.

- Automated regression testing helps implement **continuous integration (CI)**, which saves time and costs.

- It allows for the detection of defects caused by changes in the system environment.

- It increases client trust and satisfaction, which may lead to expanding business.

Smoke testing

Smoke testing is a technique that was adopted in the plumbing industry, where they usually used white smoke to identify any leaks in pipes.

Today, the concept of smoke testing is used in software development to verify the basic functionality of a build. If a test fails, the build is considered unstable, and the system is not ready to perform any other type of testing activity.

The following diagram shows the main stages of smoke testing:

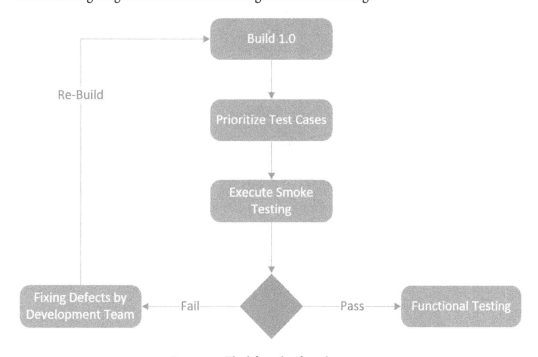

Figure 8.5: The life cycle of smoke testing

In the preceding diagram, we can see that the testing procedure starts by creating a new build with a version number. After that, we need to prioritize the test cases and decide what to test exactly to certify the new build before moving to functional testing. If the smoke testing fails, then we need to fix the defects and start over by creating a new build.

Here are some benefits of smoke testing:

- It helps to detect show-stopping issues in the early stages before starting any other type of testing.

- It improves the efficiency of the QA team by detecting defects that may take longer to be detected if they want to run functional testing.

End-to-end testing

End-to-end (**E2E**) **testing** is considered the full-fledged testing of an application. It is typically convenient to test the functionalities of the entire system; it is important to replicate the production environment to conduct this type of testing, and the testing scenarios should imitate the user behavior. The main goal of this type of testing is to certify that the different user flows are functioning properly with no errors and as per the requirements.

In the following diagram, we show the three main steps of the E2E testing process:

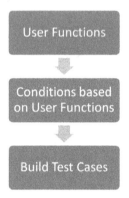

Figure 8.6: The three main steps of E2E testing

In the preceding diagram, the user functions represent the actions performed in a particular functionality in the system, and the conditions represent the various input data and sequences that can be applied to each user function. As for the test cases, these are created based on the previous two actions—that is, the user functions along with the conditions.

Here are some major benefits of E2E testing:

- It helps ensure complete readiness and the health of the system.
- It allows us to test the full system from a user's perspective.
- It helps to test real-life scenarios that can be applied by end users.

User interface testing

The term **user interface** (**UI**) speaks for itself. **UI testing** is performed to test an application's **graphical user interface** (**GUI**), with the aim of making sure that the UI of the application is developed as per the requirements and is user-friendly.

In the following diagram, we can see that the business layer and the data layer can be tested using unit tests. As for the UI, the only way to test it is through UI testing:

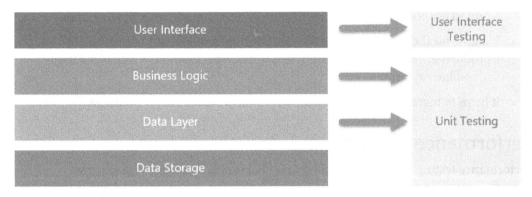

Figure 8.7: UI testing

Here are some major benefits of UI testing:

- It helps to check the alignment of UI elements, along with checking the font style, the color, and the clarity of the text displayed.

- It allows us to check whether a product is rendering correctly on all devices and screens that are supposed to be supported.

- It helps to check error messages, along with warning messages.

Acceptance testing

Acceptance testing (also known as **user acceptance testing**, or **UAT**) is considered the last phase of testing and is usually performed by the key users of the client to verify that all business requirements have been developed and that the system is working properly and efficiently as expected by the end users. Typically, acceptance testing is conducted based on test cases that are generated from user cases prepared during the analysis phase of a project.

In the following diagram, we show that UAT is the last testing phase before moving to a production environment:

Figure 8.8: UAT in the project life cycle

Here are some benefits of UAT:

- It helps to validate that all business requirements defined at the beginning of a project are correctly implemented and working properly without any errors.

- It allows for the fixing of detected defects during development rather than in a production environment, which is less costly, especially in the case of solutions with online payment.

- It helps to increase users' trust in the new system before the go-live stage.

Performance testing

Performance testing is non-functional testing that is often used to check whether a system is working properly as per the performance requirements defined by the client and the standards.

The following four main elements are considered when carrying out performance testing:

- **Bottlenecks** are major issues that bring down a system.

- The **load time** needed to load a page or a form.

- The **response time** of triggering an action or completing a process.

- **Scalability** is the ability of a system to handle a large number of requests without crashing.

The following diagram shows us the four main elements of carrying out performance testing:

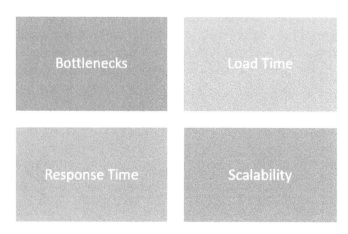

Figure 8.9: Performance-testing elements

Here are some benefits of performance testing:

- It helps to measure the response time, accuracy, and stability of the system.
- It allows for the detection of issues that reduce the response time of the application or the overall hardware usage.
- It helps improve the load time of pages and increases user satisfaction.

Stress testing

Stress testing is a type of non-functional testing that certifies the stability and reliability of a system. The main target of stress testing is to measure the strength and error-handling capabilities of the system when it is under an extremely heavy load of requests that is way beyond the normal operating situation of the system. Its purpose is to understand how the system behaves under this heavy load.

The following diagram describes the steps of stress testing:

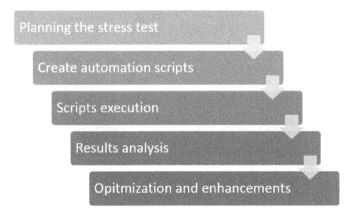

Figure 8.10: Stress-testing main steps

Initially, the stress-testing process starts by planning and deciding on the test cases. After that, we should create scripts and execute them in an automated process. The results of the stress test should be analyzed carefully to identify the root cause of any issues. At the end, we need to fix issues by optimizing the code and then rerunning the whole stress-testing process until we have a stable build.

Here are some benefits of stress testing:

- It allows us to check and handle error messages that may occur.
- It helps to check whether the data was saved correctly before any failure was caused by a heavy load of requests.

Compliance testing

Compliance testing (also known as **conformance testing**) is a type of audit-testing technique usually performed to verify whether a product meets a set of external and internal standards before deciding whether the system is ready to be released or not.

The internal standards are typically set by the organization. For example, a website should be designed for various devices and screens, therefore it should provide a responsive UI.

As for external standards, these are regulations that are set by a worldwide consortium or a third-party organization that specializes in this type of testing. An example of an external standard is the **General Data Protection Regulation (GDPR)** or the **Web Content Accessibility Guidelines (WCAG)**.

The following diagram shows the main system attributes that are usually assessed by compliance testing:

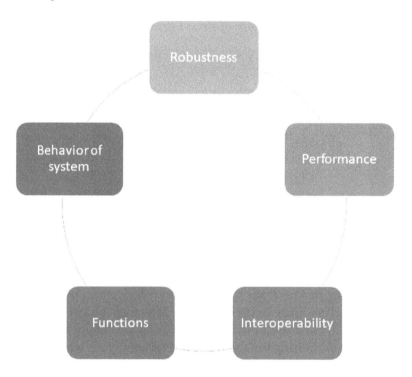

Figure 8.11: System attributes assessed by compliance testing

As shown in the preceding diagram, each attribute is contributing to the overall compliance of the system. So, let's get to know each of these attributes, as follows:

- **Robustness**: This shows the ability of a system to function normally in the case of disturbance.

- **Performance**: This represents the time needed by a system to complete a single task. Compliance testing should measure the performance of the main functions in the system and certify that they are performing well, based on predefined testing criteria.

- **Interoperability**: This shows the ability of a system to exchange information with other third-party systems. Moreover, it shows how well different functions in the system are interacting together to complete a process.

- **Functions**: This assesses the interfaces and functionalities provided by a system, along with confirming whether the requirements defined at the early stages of the project are met.

- **Behavior of system**: This assesses the behavior of a system with the environment in which it is hosted. It also assesses how the system behaves after executing every user story defined previously.

Disaster recovery testing

A **disaster recovery plan** (**DRP**) should be considered for enterprise solutions and mission-critical systems. It consists of a set of detailed guidelines and strategies that should be implemented to handle unplanned incidents that would disrupt the normal operations of a system. A good DRP should enable us to recover quickly from disruptive events such as cyber-attacks, power outages, hardware outages, or any other incidents. It should ensure the continuity of business processes and minimize damage as much as possible.

DR testing is the process of certifying a DRP by evaluating each step in the process to make sure that it will work as expected when an incident occurs.

So far, we have explored the main testing types and techniques, such as unit testing, smoke testing, performance testing, and acceptance testing. It is essential to know each of these testing types and when to use them to deliver high-quality software products. We should ensure that a product meets standards and requirements, all the way from coding to business functionalities of the product as a whole. Applying different testing types between functional and non-functional tests will boost quality, to achieve exceptional results.

In the next section, we are going to explore the capabilities of test plans in Azure.

Exploring testing in Azure

Manual testing can be a key testing technique to deliver a great UX and to certify a product before the go-live stage. **Azure Test Plans**, along with **Visual Studio 2019**, offers the features we need to manage our testing efforts, from manual and exploratory testing to load and stress testing.

The starting point is to create a test plan made up of configurations, test suites, and test cases that can be broken down into shared test steps, and use the parameters that will allow us to repeat a test but with different input data.

Use the following link to sign in to Azure DevOps: `https://azure.microsoft.com/en-us/services/devops/`.

After successful login, you can see **Test Plans** in the menu on the left side, as per the following screenshot:

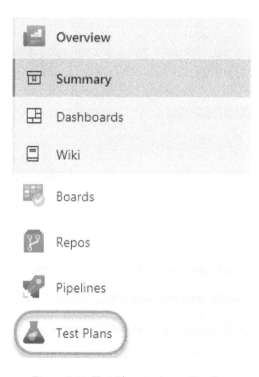

Figure 8.12: Test Plans in Azure DevOps

After creating a test plan, we need to set the configurations upon which we intend to run our tests—for example, we can specify the operating system and browser configurations if we are testing a web application. Test configurations can be assigned to an entire test plan or individual test suites, and even test cases. If we assigned the configurations to a test plan, this would ensure that all created test cases are automatically assigned to those configurations.

When you click on **Test Plans**, a sliding submenu will be opened, showing more capabilities where we can create new test plans, set parameters, and modify configurations, as shown in the following screenshot:

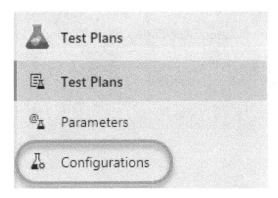

Figure 8.13: Configurations under Test Plans

With our test configurations set, we can now start creating test suites, which are collections of test cases.

There are three different types of test suites, outlined as follows:

- **Static test suite**: This is a logical container where we can add any test case we like.

- **Requirement-based test suite**: This is where we associate our test case to a work item to define its acceptance criteria.

- **Query-based test suite**: This is where we create a work-item query to select which test cases to include. Any test case that meets the query criteria will be added automatically to the test suite.

After we define our test suite, we need to start creating a test case and assigning it to the test team. Creating a test case is a very straightforward process. The main element is the steps to execute in any expected results. Steps that are repeated often can be extracted as shared steps to ease test maintenance. After preparing the test cases, we are ready to start the test run manually. Note that from the **Runs** page, we can review all our previous test runs, along with their results.

Up to now, we have learned about manual testing in Azure, which has its benefits. But when we develop more features and our source code grows in size, testing all functionalities manually can become repetitive and time-consuming. Therefore, Azure offers us a mechanism to automate our tests in order to eliminate the burden of manual testing and to allow QA engineers to focus on delivering better quality and an improved **user experience** (**UX**).

With Azure DevOps, we can automate our tests from Azure Test Plans by using Azure pipelines. There are many types of testing that we can automate with Azure pipelines, such as unit testing, security testing, and code-coverage testing, which calculates the percentage of code that's covered by unit tests.

Here are some key recommendations when using Azure Test Plans for testing:

- Make sure the testing is serving a purpose and has a positive impact on the product, and try not to test for the sake of testing.

- Keep the tests straightforward, focused, and short. Tests should run quickly, especially if they are triggered on the build or release of a product.

Summary

In this chapter, we explored some key principles that outline fundamental guidelines required to conduct proper testing. We also learned about the essential testing types that we must know as solution architects. Being aware of each of these testing types will help us decide which functional and non-functional tests we should apply to ensure high-quality software products and to deliver products that meet requirements. In the last section of this chapter, we explored the capabilities of test plans in Azure DevOps, along with the key benefits.

In the next chapter, we will dig deep into architecting modern web applications with **ASP.NET Core** and **Microsoft Azure**.

Section 3: Architecting Modern Web Solutions with DevOps Solutions

In this section, we will have an overview of the modern web solution characteristics and we will learn how to choose between traditional web apps and **Single-Page Apps** (**SPAs**). After that, we will explore the project structure in popular SPA frameworks. Then, we will explore hosting options in Azure with high-level recommendations.

Later in this section, we will get to know how to make use of Azure DevOps to build, test, and deploy our applications by using modern software development practices such as Azure Artifacts and the CI/CD practices.

This section comprises the following chapters:

- *Chapter 9, Architecting Modern Web Solutions with ASP.NET Core and Azure*
- *Chapter 10, Designing and Implementing Microsoft DevOps Solutions*

9

Architecting Modern Web Solutions with ASP.NET Core and Azure

Building rich and dynamic web solutions with **ASP.NET Core** and hosting them in **Azure** offers greater value over the traditional approach to web development practices.

This chapter provides us with a foundational understanding of how to architect web solutions with modern .NET technologies and cloud hosting scenarios.

In this chapter, we will cover the following topics:

- An overview of modern web solution characteristics
- Learning how to choose between traditional web apps and **Single-Page Apps (SPAs)**
- Understanding the project structure in the popular SPA frameworks
- Exploring hosting options in Azure with high-level recommendations

By the end of this chapter, we will have learned how to architect cross-platform modern web solutions with ASP.NET Core to take advantage of its improved performance, which is one of the most obvious benefits of this framework along with its cloud-based development support.

Moreover, we will get to know how to choose between traditional web apps and SPAs along with a quick comparison of **Angular**, **React**, and **Vue**. We will also learn how to choose the best Azure hosting approach for our solution.

Exploring the characteristics of modern web solutions

Irrespective of the industry or business of the clients, the user expectations from modern web solutions are increasing with time. End users expect to use responsive web solutions that can be accessed from different devices with various screen sizes.

Moreover, the web solutions must be secure, flexible, and scalable to allow adding new features within a short time and with less effort. Modern web solutions are expected to be easy to use with a well-developed user experience. This offers our clients a unique competitive advantage to retain their customers and distinguish themselves from their competitors. In this section, we are going to highlight the main characteristics of modern web solutions.

Scalable and cloud-hosted solutions

In the current modern era, cloud adoption is a way to accelerate digital transformation for many reasons, such as the ability to automatically scale up or down the allocation of resources based on emerging needs. Moreover, cloud hosting offers various tools to automate business operations along with strong security measures that ensure the protection of personal data and customer information that might be associated with the web solution.

ASP.NET Core is the best option, dealing perfectly with these factors. It is a cross-platform web framework that is optimized for cloud solutions. It is developed with performance and scalability in mind, which means less RAM and CPU consumption, and this will save us costs in infrastructure and hosting.

Modular and loosely-coupled architecture

Modular architecture is a design approach that consists of assembling multiple modules to construct a system. The main benefits of the modular concept are flexibility, which allows us to easily bring additional features to the system, and loosely coupled modules, which allow reducing the costs of maintenance.

It is worthwhile to use ASP.NET Core to implement the modular concept in modern web solutions. It is an open-source framework that is developed out of different **NuGet** packages. This means our web solution will only compile packages that are really needed in the solution, and it won't include additional libraries that are never used, as is the case with .NET Framework. By eliminating the libraries that are not needed in the solution, we reduce security vulnerabilities in one way or another.

ASP.NET Core is designed to allow for **dependency injection**. This is a design technique used to reduce the dependency problems between components or classes through the use of an interface, or by injecting the concrete implementation of a low-level class into a higher one.

Check out the Microsoft documentation for more information about dependency injection: `https://docs.microsoft.com/en-us/aspnet/core/fundamentals/dependency-injection?view=aspnetcore-5.0`.

Automated testing

Testing is an essential phase to certify the product we are developing. While manual testing is still important for many reasons, such as exploratory testing and visual testing, automated tests offer great benefits such as saved costs, increased productivity, high-quality products, and better performance.

ASP.NET Core allows us to easily test the system we are developing because the framework is flexible and reliable, allowing fast automated testing. It provides capabilities to easily write unit tests for **Model-View-Controller** (**MVC**) **apps** as well as **Web APIs** that are mandatory for modern web solutions. It is seamlessly integrated with Azure allowing us to have full access to the latest features in **DevOps** testing tools, which is very valuable to the product and the development team.

Traditional and single-page application support

SPAs are trendy in the web world, though it doesn't mean that every web solution should be developed using this technique. Traditional web solutions based on the MVC framework are still in demand and can be used in many cases. Traditional web solutions with **ASP. NET MVC** rely on the server to deal with the requests and render back the views, while SPAs rely heavily on Web APIs to get the data needed to render the components.

Many web solutions, nowadays, involve both the behaviors of traditional web apps and SPAs. ASP.NET supports having an MVC web application along with a web API in the same Visual Studio project. Moreover, it allows building web apps using any of the modern frontend frameworks, such as Angular, React, and Vue, along with a server-side backend web API.

Fast deployment

It is essential to easily deploy new changes to web solutions. With **Azure DevOps** pipelines, we can automate the deployment process of ASP.NET Core solutions as part of the **continuous integration** and **continuous delivery** pipeline. Microsoft Azure is also integrated with **Git** repositories, allowing the automatic deployment of new changes that are made to a particular Git branch or tag.

Moreover, we can use the tools and practices provided by GitHub, which are fully integrated with Azure, to deliver our products faster. Through GitHub Actions, which are similar to Azure pipelines, we can automate software development processes with the usage of workflows that are made up of steps and jobs. These workflows can help us build, test, package, release, and deploy any project on GitHub with an automated workflow.

For more information about the available GitHub actions for Azure, check the following reference:

```
https://docs.microsoft.com/en-us/azure/developer/github/
github-actions
```

Progressive web apps with Blazor

Blazor is a web framework that provides awesome capabilities to build interactive web applications using **C#** instead of **JavaScript**. It relies on open web standards with no need to install any kind of plugin. It can be used to build SPAs as well as **Progressive Web Applications** (**PWAs**).

PWAs are web apps that make use of the latest technologies of the browser to deliver a user experience that is similar to mobile apps. They are a powerful trend in mobile and web development. Blazor **WebAssembly** is the client-side framework that can be used to build PWA apps. Here are the benefits of this technique:

- It allows for seamless offline operations and the app can load instantly. Later, it can sync with the server to send back the data.

- Low development costs because we don't have to build different versions for multiple devices.

- It provides users with a similar UI/UX to mobile apps.

- The possibility to push notifications from the server like with native apps, even when users are not using the app.

- No need to publish the app to a store for distribution and discovery. The app can be accessed through a link or a shortcut link that can be placed in the Start menu or on the home screen.

In this section, we highlighted a set of key characteristics of modern web solutions. In the next section, we will learn how to choose between traditional web apps and SPAs.

Choosing between traditional web apps and single-page apps

So far, we have seen that there are two approaches to building web applications. One approach is the traditional way, where all the application logic is served on the server side. The other one is the modern approach represented by SPAs, where all the user interaction is handled by the browser using a client-side framework that communicates with the web server by consuming a web API. There is also a way to have a hybrid solution by combining the two approaches together in one solution.

The following diagram shows the two approaches. We can see that in the **Single Page Application**, we have multiple templates that will be rendered in one single page using a client-side framework; also, there is no full-page refresh in this approach. While in the **Traditional Web Application**, we can see multiple pages that enforce a full refresh of the page when navigating from one page to another:

Figure 9.1: Single-page application versus traditional web application

A question that usually comes to mind every time we want to architect and develop a new web solution is, *which approach should we adopt – traditional or single page?* Let's learn how to choose between these two approaches.

Selecting traditional web applications

Before we start discussing the key reasons for choosing traditional web applications, let's understand the page lifecycle of this approach. Here is a diagram showing the request lifecycle in a traditional web application:

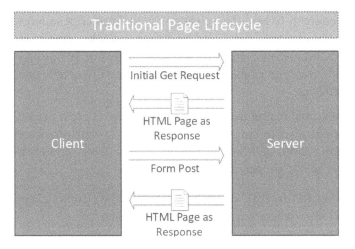

Figure 9.2: Traditional page lifecycle

In the preceding diagram, we can see the initial request is made by the user to browse a page. This request is received by the server, which will process it and return an HTML page as a response, which is considered as a full-page refresh. It is the same behavior when we post a form to the server.

A good example of this approach is the classic ASP.NET MVC application that is not making use of any JavaScript framework to render views through **AJAX** requests.

Now, let's get to know when we should choose the traditional web application approach:

- If the client-side requirements of the application are simple, then the traditional approach is a good fit. For example, most **CMS** websites are used by users to read content with a minor need for client-side functionalities. In this case, the traditional approach should be adopted to develop such applications where the actual logic is executed on the server and the response is returned as HTML to the user's browser. Check the *New York Times* website – you will notice that when you navigate from one article to another, the URL in the browser changes, which is a sign that this website is built using the traditional approach.

- If JavaScript and famous frontend frameworks such as Angular, React, and Vue have not been adopted by our team and there is not enough time to train them before we start a project.

- If the client request is to load the web app without JavaScript support, in this case, all JavaScript libraries will be disabled by default in the browser. In most cases, this is requested in intranet web apps and not online websites. In online websites, it is a must to have JavaScript enabled, otherwise, we won't be able to open the website from various devices.

- If SEO is an essential matter in the project to improve content marketing and drive more leads and readers to the website. It is possible to configure the SPA to improve the SEO ranking of the application. However, this ranking shows better results with multiple pages having proper URLs.

These are the main key factors that will lead us to choose the traditional approach. In the next section, we will learn when to choose the SPA approach.

Choosing single-page applications

An SPA is a one-page application with multiple views that are rendered using JavaScript on a single page. In the following diagram, we will explore the request lifecycle in an SPA:

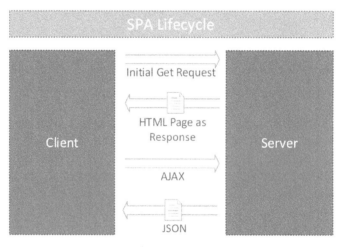

Figure 9.3: Single-page application lifecycle

In the preceding diagram, we can see the initial request is triggered by the end user opening the SPA app for the first time. The server will answer by returning the full HTML of the home page. Moreover, the user will trigger another functionality such as updating data in the database and refreshing the view.

This will be achieved through an **AJAX** technique that is used by most of the famous frontend frameworks. The **AJAX** call will consume a web service or a web API and return a **JSON** object, then it will refresh the view only without having a full-page refresh. This creates a fluid user experience allowing users to feel like they are using a native app.

Now, let's get to know when we should choose the SPA approach:

- If it is requested to provide users with the ability to work offline when they are not connected to the network or the internet. The SPA approach will give users the ability to sync their data with the server when the application is connected to the network again.

- If consuming less bandwidth is essential. It is known that SPA apps load their resources once during the initial request and they consume less bandwidth than traditional web apps because they do not load and transmit the full HTML page on every request.

- If the response time and user experience are crucial for the client. It is well known that the response time of requests in SPAs is way better than traditional applications. Moreover, the seamless and rich user experience can significantly affect the business of our clients and eventually increase leads and sales.

- If SEO is not important for the web application and if your team is knowledgeable with JavaScript, TypeScript, and any of the frontend frameworks such as Angular, React, Vue, or Blazor WebAssembly.

In this section, we learned how to choose between traditional web applications and SPA web applications. We also explained the request lifecycle for both approaches. In the next section, we are going to have a quick overview of some common SPA architectures.

Understanding the structure of SPAs with ASP.NET Core

With proper architecture, features can easily be developed, and we can reach an outstanding client's satisfaction. This approach is challenging because it requires solid architectural expertise and a proper hosting approach, but it always succeeds in delivering a decent solution.

In this section, we will get to know the structure of SPAs with ASP.NET Core. We will explore the project structure of the three top modern SPA technologies: Angular, React, and Vue.

Angular SPAs

Angular offers a full MVC pattern implementation. It is still one of the best JavaScript frameworks that is used to build SPAs. Today, with the latest version of **Visual Studio**, we can create a new ASP.NET web application with Angular enabled.

The project will be a combination of the Angular `ClientApp` folder along with the web API that is usually included in the `Controllers` folder, as shown in the following screenshot:

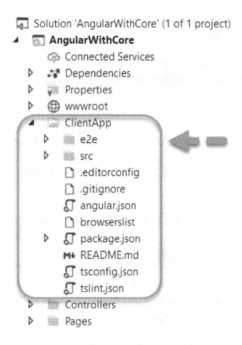

Figure 9.4: Structure of an Angular app with ASP.NET Core

The `ClientApp` folder usually contains all the files related to an Angular CLI-based frontend application and the `Controllers` folder contains all the API endpoints. The following list explains the main files and folders under the Angular client app shown in the preceding screenshot:

- `e2e`: This folder is used to create the different types of testing and it relies on a testing library called `Protractor`.

- `src`: This folder contains the frontend code that we will develop to render the components; we will spend most of the time writing code in this folder. It includes the styling file along with the configuration files to run the app.

- `angular.json`: This is the configuration file where we can specify the HTML starting page along with the main TypeScript file that should be executed at the beginning of the application.

- `.editorconfig`: This is the configuration file where we set the settings that should be applied by the editor when adding or modifying files in the Angular app.

- `package.json`: This file contains the list of dependencies that need to be available to develop and deploy the application.

- `README.md`: This contains, by default, the basic project documentation in Markdown format.

- `tsconfig.json`: This contains the configuration needed for the TypeScript compiler.

- `tslint.json`: This contains a list of rules that should be checked by the **tslint** tool to validate the quality of the TypeScript code.

React SPAs

React is one of the most popular JavaScript libraries used to build fast and interactive SPAs. It focuses on the views part of the application, mainly the UI components, therefore it requires using additional libraries to build the entire SPA.

In Visual Studio, we can make use of the existing project templates to create a new ASP. NET Core application with React. The following screenshot shows the project structure of the React application. The `ClientApp` folder contains all the files related to React and we can see the `Controllers` folder, which holds the .NET Web API:

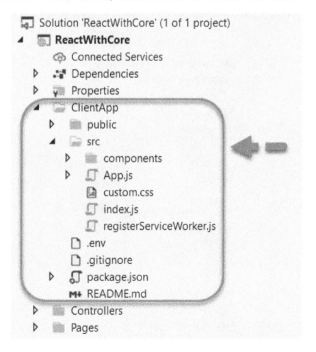

Figure 9.5: Structure of the React app with ASP.NET Core

The following list explains the main files and folders under the React client app shown in the preceding screenshot:

- `public`: This folder contains the static files of the application such as the HTML index page.

- `src`: This folder contains all the dynamic components that we will develop. It also contains the `App.js` file, which acts as the main **JS** component. As for the `index.js` file, it represents the entry point of the application that triggers the `registerServiceWorker.js` file, which is used to cache the assets of the application. This caching mechanism helps load the application faster and allows offline capabilities.

- `package.json`: This file contains the list of dependencies used in the application.

Vue SPAs

Vue is a JavaScript framework that, when combined with other libraries, is used to build modern SPAs. Unlike other monolithic frameworks, Vue is a lightweight and easy-to-learn framework. In Visual Studio, we can create an ASP.NET Core application with Vue.

Similar to the other project templates, the Vue files are included within the `ClientApp` folder and the `Controllers` folder, which contains the Web API controllers, as per the following screenshot:

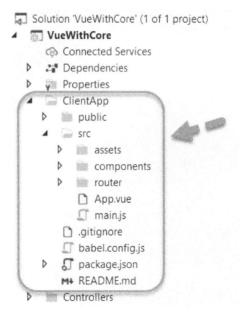

Figure 9.6: Structure of the Vue app with ASP.NET Core

The following list explains the main files and folders under the React client app shown in the preceding screenshot:

- `public`: This folder contains the static files of the application such as the HTML index page.

- `src`: This folder contains all the dynamic components that we will develop. It also contains the `App.vue` file, which acts as the root component of the application. The `main.js` JavaScript file is responsible for initializing the root component and introducing the required plugins. As for the `assets` folder, it contains all the static assets, such as the CSS files and the images.

- `package.json`: This file contains the list of dependencies used in the application.

After this quick overview of the structure of these three frameworks (Angular, React, and Vue), we may ask ourselves, *which framework should we use?* It is difficult to answer this question because it is hard to find a framework that works for every situation. Here is a table showing a quick comparison between the three frameworks:

	Angular	React	Vue
First release	2010	2013	2014
Official site	angular.io	reactjs.org	vuejs.org
Framework size	143k	97.5k	58.8k
Programming language	TypeScript	JavaScript	JavaScript
Learning curve	Steep	Moderate	Moderate
Architecture pattern	MVC framework	JavaScript library	MVVM framework
Developed by	Google	Facebook	Evan You
Used by	Google, Guardian, PayPal, Nike	Facebook, Twitter, Instagram, Uber, Netflix, Airbnb	Alibaba, GitLab, Nintendo

Figure 9.7: A quick comparison between Angular, React, and Vue

Although Angular and React are perfect frameworks to build large-scale and enterprise web solutions with complex components and very dynamic content, writing code in React is easier and faster than Angular. According to a Stack Overflow Developer Survey conducted in 2020, React is the second most popular framework after **jQuery**: https://insights.stackoverflow.com/survey/2020#most-popular-technologies.

While Vue is lightweight and easy to learn, it demonstrates the best performance between the three frameworks. Also, the Vue development community is rising steadily compared to React and Angular.

In the next section, we will get to know the three main options to host our web applications with Azure.

Exploring Azure hosting recommendations

Azure hosting offers great hosting capabilities for every business out there, whichever sector the web application is serving. It provides a wide range of cloud services that support us to host and scale web solutions. It helps us to deal with business challenges rather than spending time focusing on the infrastructure that we need to host the solution.

There are three ways to host web applications in Azure:

- **App Service Web Apps**
- **Containers**
- **Virtual Machines (VMs)**

App Service Web Apps is the recommended hosting approach for most scenarios as it offers a fully managed **Platform as a Service (PaaS)** that is optimized in a way that lets our clients focus on their business, while Azure takes care of the required infrastructure, including scaling the application. Moreover, we can make use of Azure **Static Web Apps** to automatically deploy full-stack web apps that are built using libraries and frameworks such as Angular, React, and Vue to Azure from a code repository that can be on GitHub or Azure DevOps.

> **Important Note:**
>
> Check out the Microsoft documentation for more information about the step-by-step deployment process with Azure App Service: `https://docs.microsoft.com/en-us/learn/modules/host-a-web-app-with-azure-app-service/`. Here is another link for the same thing: `https://docs.microsoft.com/en-us/learn/paths/deploy-a-website-with-azure-app-service/`.

For applications that implement microservice architecture, it is recommended to host them using a container-based approach.

> **Important Note:**
>
> Here is a Microsoft reference link on how to deploy a container instance in Azure using the Azure portal: `https://docs.microsoft.com/en-us/azure/container-instances/container-instances-quickstart-portal`. Here is another reference link on how to deploy a container instance in Azure using the Docker CLI: `https://docs.microsoft.com/en-us/azure/container-instances/quickstart-docker-cli`.

If your application is not fully ready to be hosted on the cloud and if you would like to have more control over the hosting environment, you can choose **Azure Virtual Machines**, which is an **Infrastructure as a Service (IaaS)**. However, if you choose this option, you must take into consideration that you need an ongoing maintenance effort to manage the VM environment and keep it up to date.

> **Important Note:**
>
> Here is a reference on how to create a Windows virtual machine in the Azure portal: `https://docs.microsoft.com/en-us/azure/virtual-machines/windows/quick-create-portal`. Here is a Microsoft reference on how to create a Linux virtual machine in the Azure portal: `https://docs.microsoft.com/en-us/azure/virtual-machines/linux/quick-create-portal`. Here is another step-by-step guide on how to deploy a website with Azure virtual machines: `https://docs.microsoft.com/en-us/learn/paths/deploy-a-website-with-azure-virtual-machines/`.

Summary

In this chapter, we explored some key characteristics of modern web solutions that we must know to build scalable and cloud-hosted solutions. We learned about the difference between traditional web applications and SPAs, and how to choose between them.

Moreover, we had an overview of the project structure for the three modern frontend frameworks to build SPAs with the ASP.NET Core Web API, and we provided a quick comparison table between these frameworks. Later in this chapter, we explored the main options to host web applications in Azure with high-level recommendations to know how to choose the best hosting approach for our solution.

In the next chapter, we will dig deep into designing and implementing **Microsoft DevOps** solutions and we will learn about their benefits.

10

Designing and Implementing Microsoft DevOps Solutions

In the previous chapter, we learned about the key characteristics of modern web solutions. We also explored the project structure of the three main frontend frameworks and provided a quick comparison. We then learned the three hosting options in Azure and how to choose the best hosting approach for our solution.

In this chapter, we will learn how to effectively plan and manage **DevOps** processes while building Microsoft solutions. Azure DevOps offers a set of modern tools that allow us to plan smarter and develop a product faster. It also provides solid collaboration between the team members to deliver better quality products.

In this chapter, we will cover the following topics:

- Exploring agile planning with **Azure Boards**
- Learning about source control

- Understanding Git repositories, along with branching and pull requests
- Getting to know **Azure Artifacts**
- Understanding the logic behind the CI/CD practices

By the end of this chapter, we will have learned how to make use of DevOps to build, test, and deploy our applications using modern software development practices. Moreover, we will know about Work Items, and we will have learned about **Git** and its main capabilities. We will have also explored how to manage packages using Azure Artifacts, and also understood the **continuous integration/continuous development** (**CI/CD**) practices.

Now, let's take a look at the key characteristics of modern web solutions.

Exploring Agile planning with Azure Boards

Azure Boards is a service provided by **Microsoft** as part of the Azure DevOps service. It provides a set of features and capabilities for managing the entire life cycle of the software project. It includes tools for managing **Work Items**, **Sprints**, and **Backlogs**. Moreover, it provides end-to-end predefined and customizable dashboards, allowing us to dig deeper into the big picture of the project's activities, alongside essential **KPIs** and metrics, to understand how the project is progressing.

Let's start by exploring the core features of Azure Boards.

Introducing Work Items

Work Items are the core components in Azure DevOps and can help our Agile team manage their daily work, organize Sprints, and prioritize tasks in Backlogs. A Work Item can be a general task, an issue, or a requirement. The following screenshot shows the landing page of **Work Items**:

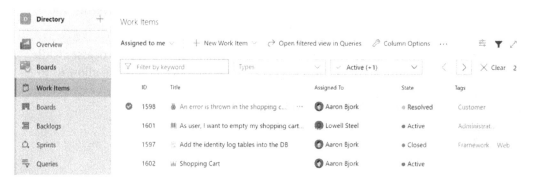

Figure 10.1: Work Items landing page

The preceding screenshot represents the home page of all **Work Items**, where we can filter items based on specific criteria. We can also assign items, add new items, and manage existing ones. This page provides every person that's working on the project with a complete view of the progress, along with the status of each item and who is doing what. We can filter to see tasks that were planned to be delivered in the next **Sprints**.

We can also specify the dependencies between the items to break large tasks down into smaller, more manageable items, as well as create queries and save them for later use. A query is a filtered view of all the **Work Items**. For example, we can create a query to display the active tasks, or a query to display tasks that have been assigned to a particular team member.

It is easy to create a new Work Item. As shown in the following screenshot, we just need to click on **New Work Item** and then choose the type; that is, **Epic**, **Issue**, or **Task**:

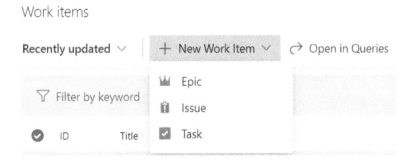

Figure 10.2: Action menu for creating a new Work Item

In the action menu, we can see three main types of Work Items:

- **Epic**: This represents a large item that's required for the product to function. It can be broken down into smaller user stories. A user story is a specific Work Item within **Epic**. For example, let's assume we have received a request to implement a login mechanism for an e-commerce website. In this case, the Epic represents this request. The user stories here could be **Login with Email**, **Login with Google**, **Login with Facebook**, and **Forgot password**.

- **Feature**: This represents the bulk of the functionality that fulfils users' needs. A **Feature** is a collection of user stories that delivers business value and the context of the software product.

- **User Story**: This represents the smallest element in the Agile methodology and describes a requirement or a need from a user perspective. To create a **User Story**, we should follow the `role-feature-benefit` template: as a (*user role*), I want (*an action/or goal*) so that (*a benefit/or reason*); for example: as a (*customer*), I want (*a shopping cart functionality*) so that (*I can buy items and pay online*).

- **Issue**: This represents bugs, code defects, and software issues that we may capture while developing the product.

- **Task**: This represents a Work Item that has been planned as part of building the product. This can be either a result of an issue or requirements, including requirement analysis, development, or testing.

The following screenshot shows the details page of a sample Work Item:

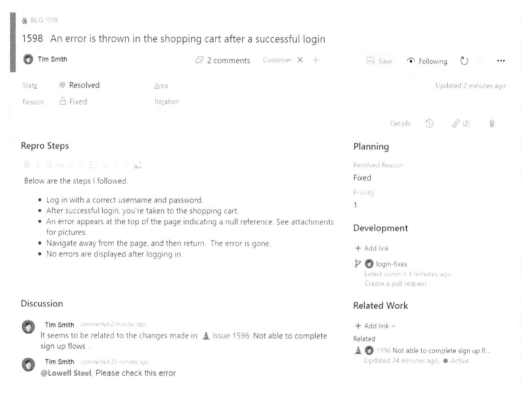

Figure 10.3: A bug item details page

On the details page, we can see that every Work Item has a title with a unique ID, status, and iteration, along with the steps to reproduce if it is a defect, or an item description if it is a requirement.

We can also see the comments that are attached to the Work Item. These represent the discussion happening between the team members about this Work Item. We can follow a Work Item to receive notifications whenever there is an update. We can also assign it to a team member, as well as link it to another Work Item by, for example, linking an issue to a task or Epic.

In the next section, we will learn how to use Work Items to report and organize work.

Exploring Boards, Backlogs, and Sprints

In the previous section, we learned about Work Items, so let's learn how to use them in Boards, Backlogs, and Sprints to organize and track team deliverables.

The following screenshot shows a sample board that was associated with a project upon its creation:

Figure 10.4: Sample Kanban Board

Every time we create a new project, there is a preconfigured **Kanban Board** that is created and linked to the project so that we can visualize the progress of work items. This board is fully customizable. We can drag and drop items from one category to another to reflect the current situation of the project. We can also organize tasks by status, ownership, Sprints, or any other criteria.

Backlogs help us keep things organized according to priorities. As shown in the following screenshot, Backlogs provide a flat view of the Work Items, while Boards display them as cards:

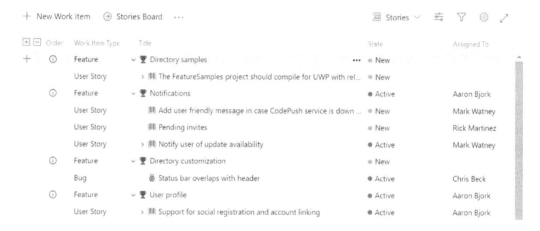

Figure 10.5: Backlogs list view

The product backlog should reflect the plan and roadmap of what we plan to deliver in the upcoming Sprints.

Finally, **Sprints** are the heartbeat of DevOps as they represent the iterations of an Agile project. A Sprint has its own **Capacity** planning and **Taskboard**. It should be short in terms of duration, typically between 1 to 4 weeks; during this period, there must be a set of Work Items that should be accomplished by the team. The following screenshot shows a sample **Sprint** view:

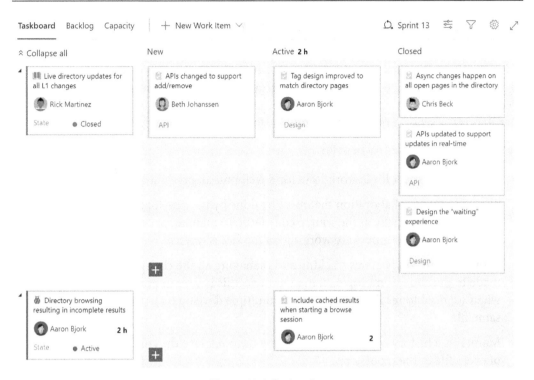

Figure 10.6: Sprint view

In the preceding screenshot, we can see how the tasks are organized in the **Taskboard** area to reflect the plan of a **Sprint**. We can drag and drop items between the different categories, and we can check the overall progress of the team in this **Sprint**.

> **Important Note:**
>
> The Microsoft documentation for learning about and understanding everything related to Azure Boards can be found at `https://docs.microsoft.com/en-us/azure/devops/boards/?view=azure-devops`.

In this section, we learned how to define the project roadmap and plan Work Items. This helps our team break down complex solutions into manageable workloads by using a robust platform from Azure DevOps. In the next section, we will learn about source control in Azure DevOps.

Getting started with source control

Source control (also known as **version control**) is an essential part of DevOps. It is a collaboration platform that can be used by the development team to track and manage changes in the source code. It provides a historical version of each source code file in the project. It also helps resolve conflicts when merging changes from multiple developers. **Azure Repos** is a set of version control tools that we can use to manage our code.

Here is a list of source control benefits:

- Ability to create multiple workflows for development, production, and testing.

- There is a lot of collaboration that must be done by the development team to deliver the product, especially at the source code level, to maintain a common repository when multiple developers are working on the same project.

- Source control supports us tracking and managing all the changes that have been made to the code by multiple team members. This is very important, especially when we need to resolve conflicts when multiple developers try to modify the same file.

- Maintains a history of changes by allowing us to retrieve the complete history of every file in the repository.

- Ability to label the source code to keep track of the product version, especially when we have multiple releases. We can also create branches to manage the development activities between the production and development environments.

Azure Repos provides two types of version control:

- **Team Foundation Version Control** (**TFVC**): The code history is centralized on the server and team members need to be connected to check in.

- **Git**: The code history is distributed on each team member's machine, where they can commit changes locally.

Scaling Git for enterprise DevOps

Git is one of the most essential version control systems that is adopted by development teams and companies. Git is a distributed version control system, which means the local copy of the source code that's stored on each machine represents a complete version control repository.

In this section, we will learn more about Git and how to structure repositories, manage branches, and collaborate with pull requests.

Structuring Git repos

There are two types of repositories that we can use with Git:

- **Mono-repo**: More than one project is stored in a single repository
- **Multi-repo**: Each project has its own repository

Mono versus multi; *what's the right approach?* There is no direct answer that would recommend a particular approach. The strategy that we choose to use in order to structure our repositories is totally based on our way of managing projects; both types have their advantages and disadvantages. However, it is good to mention that **Facebook** and **Google** use mono-repos to manage their projects. Here are some key points to help you decide which strategy to follow:

- Mono-repo facilitates managing dependencies that may be complex if we use multi-repo.

- With mono-repo, we may face some performance drawbacks in the case of a large code base. This is not a problem in multi-repo.

- It's noticeable that enforcing common practices and standards is easier in mono-repo than in multi-repo.

- Multi-repo allows us to work efficiently by enabling each microservice team to work independently to finish their work faster. This allows us to grant developers access to the repositories they need to access.

Branching strategy with Git

Git branches are effective references to a snapshot of the code changes. A branch provides a way to isolate changes related to a new feature or a hotfix from the main branch of the code. Code changes that are committed to one branch don't affect the other branches automatically, unless we merge changes.

It is essential to adopt a branching strategy and make it simple by following these three concepts:

- Create a new branch for every feature or set of features of a particular release. This is also applicable in the case of releasing hotfixes after fixing defects.

- Merge sub-branches into the main branch by using pull requests. Never merge code into the main branch unless the code had been tested properly, and also ensure that the affected functionalities are working well and certified.

- Keep the main branch up to date and never modify code directly inside it.

The following diagram shows how all the sub-branches merge their updates into the main branch:

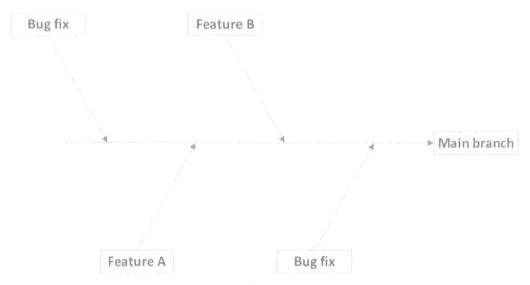

Figure 10.7: Merging sub-branches into the main branch

There are many branching strategies that you can implement. The most important part is to separate the development activities from the production code by creating two separate branches. One of the strategies that we can follow is the **trunk-based branching** strategy, as shown in the following diagram:

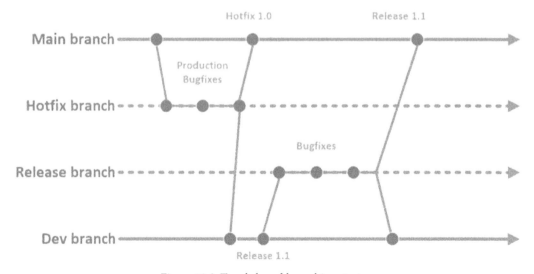

Figure 10.8: Trunk-based branching strategy

In the preceding diagram, we can see the two main branches: development (**Dev**) and production (**Main**). The concept here is that we never write code directly into the **Main** branch. Instead, we need to create a branch for hotfixes; at the same time, the hotfixes should be merged with **Dev** after proper testing.

As for the **Release** branch, it is usually created from the development branch. After development and proper testing, it is merged with both the production and development branches. This way, we make sure that the **Main** branch contains the production version of the code, while the **Dev** branch contains the development branches.

Git branches are inexpensive to create and maintain. Therefore, as shown in the preceding diagram, we created a separate branch. Even small fixes and changes should have their own feature branches, which should simplify reviewing the history of the changes. When creating a new branch, it is important to provide descriptive information about the branch and link it to a Work Item.

Collaborating with pull requests in Azure repos

Pull requests are robust mechanisms for notifying the team leader or the code reviewer that the development of a new feature or a bug fix has been completed, and that the code must be reviewed before it's merged into the main branch. Avoid merging code to the main branch without a pull request, which enforces a code review process. This is essential for improving the code's quality.

It is noticeable that if the feedback that was received after the review process is good and up to standards, it may improve the code's quality. Therefore, it is recommended that you provide high-quality feedback. Here are some key suggestions for successful pull requests:

- Having the right people to review the pull request and provide feedback is a key factor for better reviews.

- It is recommended to have two reviewers as an optimal number for the review process.

- Giving actionable and constructive feedback is very essential.

- It is important to reply to comments promptly to accelerate the pull request process, especially if you have a large number of requests in the queue.

- Providing enough details in the branch description helps the reviewer understand the purpose of the changes.

- It is recommended to combine the code review sessions, if you have them in place, with the pull request process to avoid duplicating the effort.

In this section, we learned about the structuring options that we can use in Git repositories and how to choose between them. We also explored some recommendations for a better branching strategy and discussed the benefits of the pull request process. We then highlighted some key factors for improving this process. In the next section, we are going to learn how to set a good dependency management strategy.

Managing packages with Azure Artifacts

Azure Artifacts is an extension in Azure DevOps that provides a set of capabilities to create and manage packages with **NuGet**, **npm**, and **Maven**. This can help us manage the dependencies in our code base and group them into feeds. Each feed that's created in Azure Artifacts has its own URL that we can consume from **Visual Studio** to install the packages into our solution; the development team can also use the same feed URL to publish private packages.

Azure Artifacts is free as long as the size of the packages and artifacts is less than **2 GB**. Everything above 2 GB will be billed according to the subscription plan. The following page on Azure Artifacts can be found in the left-hand side menu, next to the arrow depicted in the following screenshot:

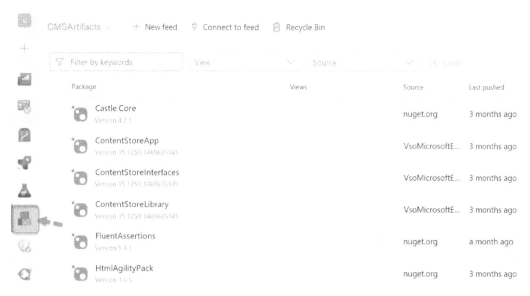

Figure 10.9: Packages feed within Azure Artifacts

In the preceding screenshot, we have a feed called **CMSArtifacts**. In this feed, we can see a group of packages that were added to this container. The purpose here is to organize the public and private packages that we are using in our solutions into a feed that can be consumed by the development team.

With Azure Artifacts, we can create views of the feed. For instance, we can create a view for the packages that are used in the development environment and another view for the production version of the product.

The following screenshot shows three different views of the same feed; that is, **Local**, **Prerelease**, and **Release**. Each view holds a particular version of the packages, and it is being used for a particular work environment:

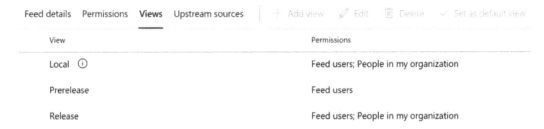

View	Permissions
Local ⓘ	Feed users; People in my organization
Prerelease	Feed users
Release	Feed users; People in my organization

Feed details   Permissions   **Views**   Upstream sources | + Add view   ✏ Edit   🗑 Delete   ✓ Set as default view

Figure 10.10: Feed views

As we can see, there are three views in the preceding screenshot. These views were created alongside the feed. We can still add new views or modify an existing one.

Upstream source, as shown in the preceding screenshot, allows us to group the packages that we create along with the packages that we consume from the remote feeds in one place. The following screenshot shows the interface we can use to create an upstream. Notice that we can specify the type of **View** that we want to use for the upstream:

Add a new upstream source

Feed *

| CMSFeed | ⌄ |

Package type(s) *

☐ npm   ☑ NuGet   ☐ Python   ☐ Maven   ☐ UPack

View * ⬅ ▬

| Local | ⌄ |

Upstream source name *

| CMSFeed@Local |

Figure 10.11: Adding an upstream source with a specific view

Each upstream source is linked to one view, and that's how we can make use of the views in Visual Studio through upstream sources.

In this section, we introduced Azure Artifacts, which supports the multiple feeds approach. We can make use of it to organize and group the packages that we are consuming in our projects. For more technical information on how to create and manage Azure Artifacts, please refer to the Microsoft documentation: `https://docs.microsoft.com/en-us/azure/devops/artifacts/overview?view=azure-devops`.

In the next section, we will explore continuous integration with **Azure Pipelines**.

Exploring CI/CD with Azure pipelines

Continuous integration, **continuous delivery**, and **continuous deployment** (or **CI/CD**) are the main pillars of building, testing, and deploying robust applications using modern software development techniques in DevOps. These practices allow us to release new features and fixes quickly through automated processes. Let's get to know each of these practices.

Continuous integration (**CI**) is the nucleus practice in DevOps. It allows us to frequently integrate all source code modifications coming from multiple developers into the main repository. It is an automated process that can be configured in Azure DevOps. When this capability is enabled, every time a developer commits their code, the CI will be verified by starting an automated build process to verify that the project contains no build errors. After that, an automated testing process is triggered to confirm that the newly committed code is stable. This approach is very helpful for identifying errors quickly and easily.

Continuous delivery is a practice that automates the delivery step that comes after the building and testing phase. Whenever we have a successful build and tests, an automated process is triggered to deploy the artifacts to the desired environment. This can be staging or production. Note that in this practice, shipping the code from staging to production is completed through manual intervention.

Continuous deployment has a lot in common with continuous delivery. The main difference is that this practice automates the entire life cycle of the release process, and the artifacts are automatically deployed to production.

The following diagram shows the steps of each practice:

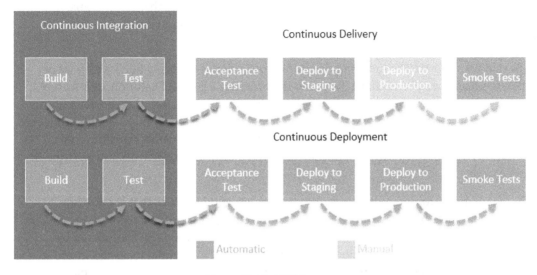

Figure 10.12: CI/CD steps

In the preceding diagram, neither the **Continuous Delivery** nor **Continuous Deployment** processes can start unless the **Continuous Integration** process is completed. The steps between **Continuous Delivery** and **Continuous Deployment** are almost the same; however, in **Continuous Delivery**, the deployment to production is done through a manual job, while in **Continuous Deployment**, it is an automatic process.

To implement a build strategy, we need to make use of the pipelines in Azure DevOps. A pipeline is an automated service that's used to verify a build and make it ready for deployment. The usage of the pipeline will reduce the manual work needed from the developer to merge the code, build it, and test the changes, along with the affected features. It is important to mention that the pipelines are used in continuous delivery and continuous deployment to automate their steps.

Summary

In this chapter, we explored the fundamental procedures of Agile planning in Azure DevOps. We also learned about Azure Boards, along with related components, such as Work Items, Backlogs, and Sprints. Then, we discussed source control and explained the difference between TFVC and Git.

After that, we explored Git and how it can version source code, before learning about branching and pull requests. Later, we learned about the packages that are available in Azure Artifacts, along with CI/CD, which help automate the steps related to building, testing, and deploying our code.

Now that you have finished reading this book, your mind is probably filled with a lot of different ideas since you've dived deep into the everyday aspects of solution architecture. I suggest that you start by measuring where you are on your journey toward becoming an effective solution architecture. A good solution architect helps build high-quality products that fit the existing environment, along with the clients' requirements. To achieve this, a solution architect must learn about each part of the business model and how these parts work together.

We covered many topics in this book. However, it is a good practice that we develop a learning mindset by frequently researching and getting to know new techniques and patterns in solution architectures, and also focus on the cloud services offered by Azure for building modern solutions. At the same time, it is essential to always improve our soft skills, especially if we want to become effective leaders. I hope that you have enjoyed reading this book as much as I enjoyed writing it and sharing my thoughts and experiences. I wish you every success in all your .NET projects!

Other Books You May Enjoy

If you enjoyed this book, you may be interested in these other books by Packt:

Enterprise Application Development with C# 9 and .NET 5

Ravindra Akella, Arun Kumar Tamirisa, Suneel Kumar Kunani, Bhupesh Guptha Muthiyalu

ISBN: 978-1-80020-944-2

- Design enterprise apps by making the most of the latest features of .NET 5
- Discover different layers of an app, such as the data layer, API layer, and web layer
- Explore end-to-end architecture, implement an enterprise web app using .NET and C# 9, and deploy the app on Azure
- Focus on the core concepts of web application development such as dependency injection, caching, logging, configuration, and authentication, and implement them in .NET 5
- Integrate the new .NET 5 health and performance check APIs with your app
- Understand how .NET 5 works and contribute to the .NET 5 platform

Software Architecture with C# 9 and .NET 5

Gabriel Baptista, Francesco Abbruzzese

ISBN: 978-1-80056-604-0

- Use different techniques to overcome real-world architectural challenges and solve design consideration issues

- Apply architectural approaches such as layered architecture, service-oriented architecture (SOA), and microservices

- Leverage tools such as containers, Docker, Kubernetes, and Blazor to manage microservices effectively

- Get up to speed with Azure tools and features for delivering global solutions

- Program and maintain Azure Functions using C# 9 and its latest features

- Understand when it is best to use test-driven development (TDD) as an approach for software development

- Write automated functional test cases

- Get the best of DevOps principles to enable CI/CD environments

Packt is searching for authors like you

If you're interested in becoming an author for Packt, please visit `authors.packtpub.com` and apply today. We have worked with thousands of developers and tech professionals, just like you, to help them share their insight with the global tech community. You can make a general application, apply for a specific hot topic that we are recruiting an author for, or submit your own idea.

Share Your Thoughts

Now you've finished *Solution Architecture with .NET*, we'd love to hear your thoughts! Scan the QR code below to go straight to the Amazon review page for this book and share your feedback or leave a review on the site that you purchased it from.

`https://packt.link/r/1-801-07562-X`

Your review is important to us and the tech community and will help us make sure we're delivering excellent quality content.

Index

www.ingramcontent.com/pod-product-compliance
Lightning Source LLC
Chambersburg PA
CBHW060547060326
40690CB00017B/3631